Do It Yourself
12 Volt
SOLAR POWER

Michel Daniek

Permanent Publications
A Simple Living Series Book

Published by:
Permanent Publications
Hyden House Limited
The Sustainability Centre
East Meon
Hampshire
GU32 1HR
Tel: 01730 823 311 or 0845 458 4150 (local rate UK only)
Fax: 01730 823 322
Overseas: (int. code +44 1730)
Email: info@permaculture.co.uk
Web: www.permaculture.co.uk

© 2007 Michel Daniek

The rights of Michel Daniek to be identified as author of this work has been asserted by him in accordance with the Copyright, Designs and Patents Act 1988.

Designed and typeset by John Adams and Tim Harland

Printed by CPI Antony Rowe Ltd
Chippenham, Wiltshire

Printed on 75% recycled paper paper

British Library Cataloguing-in-Publication Data
A catalogue record for this book is available from the British Library

ISBN 978 1 85623 039 1

Disclaimer
Everything in this book has been carfully tested by the author, but neither the author or the publisher shall have liability for any damage or loss (including, without limitation, financial loss, loss of profits, loss of business or any indirect or consequential loss), however it arises, resulting from the use of, or inability to use, the information in this book.

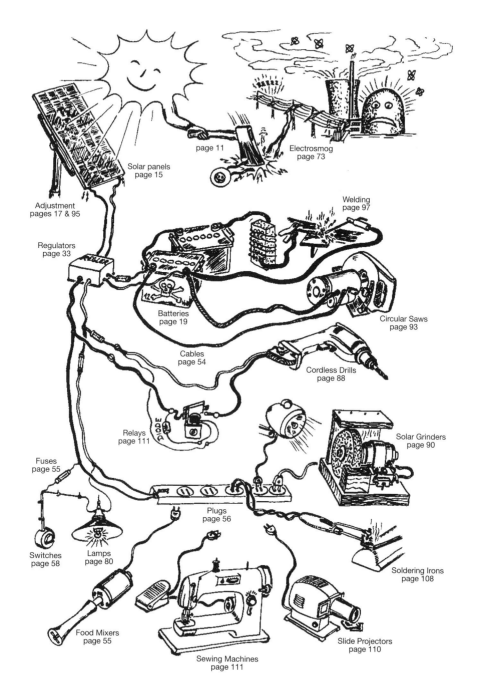

page 11

Electrosmog
page 73

Solar panels
page 15

Welding
page 97

Adjustment
pages 17 & 95

Regulators
page 33

Batteries
page 19

Circular Saws
page 93

Cables
page 54

Cordless Drills
page 88

Relays
page 111

Solar Grinders
page 90

Fuses
page 55

Plugs
page 56

Switches
page 58

Lamps
page 80

Soldering Irons
page 108

Food Mixers
page 55

Sewing Machines
page 111

Slide Projectors
page 110

And much, much more...

Contents

The Author

Michel Daniek, was born 1964 in Giessen, Germany. He grew up the only son of a plumber and from an early age enjoyed playing in his father's workshop. Whilst still a teenager he became an industrial mechanic in the motor industry but he soon started wondering if there could be alternative ways of living. He then worked as a bicycle mechanic for handicapped people for many years. But on reaching 30 he found himself totally dissatisfied with the German way of life, bought himself a truck and  left in search of other ways to live. During his travels he experimented with a small solar system in his truck and ever since has used solar energy in his day to day life. In 1997 he finally settled in a new home in an alternative village in the sunny south of Spain.

Foreword

by Ben Law

This excellent practical book contains all you need to know to set up a 12 volt, off grid solar system. It contains a wealth of information from constructing 12 volt circular saws to electric guitars. There is even a solution for making a 12 volt washing machine, the 'holy grail' amongst many of us living off the National Grid.

Michel Daniek cleverly combines his mechanical training with solar energy to offer many DIY solutions for anyone living or aspiring to live with an autonomous energy supply.

I have been living with 12 volt electricity for 16 years, and I have had to learn, change, upgrade and experiment to reach the system I now have at my Woodland House, running on 12 volt solar and wind power with a large battery storage capacity. I made many mistakes along the way, like suffering voltage loss from insufficient sized cables in my first caravan. I struggled to find the necessary knowledge and experience to help me unravel what at the time felt like complex physics.

Yet, here in one book are all the answers I needed. Volts, amps, watts and ohms are explained with logical clarity – and batteries don't wear out because Michel has solutions to repair them! This is what is so empowering about this book, it allows and encourages you to create simple 12 volt tools from everyday items such as windscreen wiper motors and motorbike starter motors. I am particularly impressed with the bicycle wheel sun following system to gain the maximum potential from your solar panels. And if you have problems with your panels, Michel has answers for how to repair them.

For anyone who is living in a truck, bender, caravan, yurt or other low impact dwelling, or if you are inquisitive about options beyond the national grid and alternative energy supplies, *DIY 12 Volt Solar Power* is full of practical solutions and is sure to become a well thumbed classic for many off griders around the globe.

Ben Law
Author, woodlander and off grid ecobuilder
October 2007

Abbreviations & Symbols

AC	Alternating current
DC	Direct current
V	Volt
W	Watt
kW	Kilowatt
kWh	Kilowatt hours
A	Ampere
Ah	Amp hours
mA	Milliamp
mAh	Milliamp hours
Ω	Ohm
kΩ	Kilo Ohms
MΩ	Mega Ohms
Hz	Hertz
kHZ	Kilo Hertz
MHz	Mega Hertz
GHz	Giga Hertz
kg	Kilogram
kg/l	Kilograms per litre
Ø	Diameter
EMF	Electromagnetic field
ESAF	Electrostatic alternating field
LED	Light emitting diode
UV	Ultra violet

About This Book

When I started to work with solar energy, I thought I had found a solution for at least one part of our global problems. Very enthusiastic, I started to give solar workshops all over Germany. I found it quite easy to convince and inspire people but after a year I was sick of telling the same thing over and over again, so I decided to write this book. It was about this time that I moved from cold Germany to this crazy town here in the south of Spain which with its abundant sunshine, has proved a perfect playground for my solar experiments. When I first arrived this area still needed a lot of development – many people were still living with candles. It was a pleasure for me to bring the luxury of electric light and music into their lives.

However energy use consciousness was sadly lacking. Things developed really quickly and the level of consumer demand soon outpaced the capacity of my small solar systems. Many people changed to using electricity from the National Grid. My idealism with solar energy was badly punished but it opened my eyes to the fundamental problem behind the destruction of our planet: consciousness.

I started to write a book about it but the theme is so huge... However I still think that small scale solar energy is an important field of learning to increase awareness of one's own energy use. It's just one little step, but it is a step in the right direction.

So... let the sunshine in!!!

Thanks to all of you! Thanks for all the great help, enthusiasm, idealism and all the love that the world needs so badly in these times of change.

This book was originally written in German in the sunny winter of 1997-98 in El Morreon, Spain. It was translated into English by Cathy Green, Rod Wilson and Abi Hill nearly at the same time. The drawings are by Carma Sola, Demian Oyarce, Michel Daniek and Marion Miller. Proof reading was done by Achim, Felix, Elke, Frederik, Christian, Günther and Patricia.

In 1999 it was translated into Spanish by Concha Buenaventura and Natalia Rodriguez. A new typed Spanish version will soon be available with great help from Nadin, Timbe-Drums and Concha Buenaventura.

In the summer of 2005, I rewrote and typed out the English version, adding some new material, including many experiences from the previous seven years of living and working with solar electricity in southern Spain. Special thanks for Jeem, Daniel Wahl and Patrick Whitefield for their good help with that.

The German version, *Einfälle statt Abfälle – Solarstrom in 12V Anlagen*, published in 1998, can be ordered quoting: ISBN 3 924038 58 9. Or you can buy it direct from:

Einfälle statt Abfälle, Christian Kuhtz, Hagebutten str.23, D-24113 Kiel, Germany. Or send a fax order to: +49 431 320 0686.

For any further questions please write to me at:

Michel Daniek, Apartado 254, E-18400 Orgiva, España. Or email: solarmichel@hotmail.com

Michel, August 2007

Introduction

We all know about the problems of CO_2 and radioactivity. Solar electricity is an important step away from the fossil fuel crisis and the potential catastrophe of nuclear energy. Solar is the energy of the future!

The number one energy source will always be the sun! However although all the techniques and materials are available, solar energy is not widely used. The technology of using energy direct from the sun is being suppressed by big business and political interests. These groups spread misinformation (see the chapter, Solar Panels, page 15) and use solar models, inventions, projects and studies for the sole purpose of proving them non-viable. They claim that solar energy is too expensive – but we should weigh this cost against the true cost of nuclear contamination and CO_2 output from fossil fuels. At the end of the day, the energy industry's only concern is simply to make as much money as possible.

With this as their aim the electricity industries produce more electricity than is actually needed, giving consumers the illusion that there is no limit to the amount of fuel they can consume. There is little consciousness of a problem and people rarely think about how much electricity they are using.

There is also the problem of 'electrosmog' – the electromagnetic field around electricity pylons and wires. High voltage electricity pylons produce an electromagnetic field with a frequency of 50 Hertz (see the chapter, Electrosmog, page 73) which pollutes large areas of land. People are becoming more aware of this problem and minimising it is now one aspect of environmental house building. However studies on damage to living organisms caused by electrosmog are either refused funding or the results are not published.

The electricity industries do not produce many power saving appliances or teach people to use energy more efficiently. Instead they put products on the market such as TV's with standby buttons and appliances that use power even when they're switched off. These appliances are continually adding to your fuel bill, CO_2 emissions and creating electromagnetic fields (electrosmog) even when they are not being used. We can stand up to these commercial and political manipulations by using alternative, decentralised energies to show that it is possible to live free from the grip of large energy concerns. We can lead by example!

Alternative energy includes fast growing biomass fuels such as rape, hemp and coppiced willow; heaters using insulated storage tanks; solar cars; wind and water energy. Last but not least is photovoltaic energy for producing solar electricity. This must become affordable for everyone and part of our everyday lives. It is important that growth is encouraged – the more consumers spend money on solar panels, the more producers will be encouraged to market them, drastically reducing the cost. But within this process there must also be a change in patterns of energy consumption because todays wasteful attitude cannot be the basis of a solar future.

For many people, solar energy is already a part of their lives. People with many different motivations are creating new power systems and houses with autonomous energy supplies. There are solar associations spreading information and setting up groups to create solar systems. However whereas a solar water heating system will pay back the initial financial outlay within only a few years, a solar electricity system connected to the National Grid is only financially viable if it's on a large scale. Therefore most solar power systems are now in alternative communities where they help to fulfil the desire for an autonomous and ecological life.

Low voltage solar power systems can allow us to maintain our standard of living without the electrosmog of the National Grid. Solar power systems with battery storage are being put together by people looking for new ways of living and a way out of this apparent dead end as we become less and less able to ignore the worldwide ecological crisis all around us. More and more people want to act instead of just sitting back and watching the earth being destroyed. Many people would like to have a future!

We are working on small solutions to our ecological and social problems and creating new ways of living and working. But the way is full of compromises and will therefore be criticized by more radical environmentalists and also by politicians who do not see solar energy as a viable large scale solution. But we shouldn't be surprised that the problems and structures of generations aren't overcome within a few days. These problems are within us all and not simply somewhere 'out there' in society. So long as we only try to change individual parts without looking at the whole and all its human causes, our search will be in vain.

Small, independent solar systems serve as examples to teach awareness of energy consumption. In terms of the world's ecological problems these systems are just a drop in the ocean and often bring with them their own ecological problems, for instance energy storage in batteries. But they are important first steps and vital areas for learning.

Please can I use your electricity for my hammer?

This book aims to provide you with a basic knowledge of solar power systems along with many tips and ideas. I will show you how to build small and medium sized solar power systems made from easily obtainable materials. There is a basic electricity lesson about volts, amperes and watts, and instructions for using a digital multimeter. I would like to show you some simple uses for solar electricity. You will be amazed by the possibilities of solar power!

Solar Panels

These are the heart of every solar power system. They are made from pure silicon crystals. Silicon is the second most common element found in the Earth. Through a special procedure these crystals acquire the property of transporting electrons when light is shone on them. This procedure is very complicated and you could fill a whole book just on this 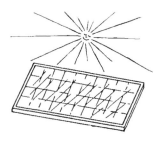 subject. There are different types of solar cells: polycrystalline, monocrystalline, triple cells and amorphous silicon cells.

Amorphous silicon cells are dark browny red, easy and cheap to make, and are used in solar calculators and watches. However they have one disadvantage – they lose power after 8 or 10 years.

Triple cells are made from three layers of silicon, each layer absorbing a different light frequency. This means that they are better than other types of cells at producing power when it's cloudy, or when they are partly in shadow. In addition they are cheap. They were an interesting development but you rarely find them. This is because they are also made from amorphous silicon and lose power too quickly, so that they might not even last the promised ten year guarantee period.

Mono and Polycrystalline cells have proved effective over many, many years. Polycrystalline cells have nice shining blue crystal patterns. Monocrystalline cells are plain dark blue and give slightly more power per cm^2 than polycrystalline cells.

 However because polycrystalline cells require less energy to produce, they take less time to recoup the energy that goes into their production. For our purposes, polycrystalline cells are the most useful.

There are the wildest rumours and scare stories about solar panels – such as claims that they are toxic. This story stems from the fact that early models were made with plastics containing Fluoride. However today they are made with PVC instead which is recyclable and non toxic. There are some toxic satellite solar cells but these are not the sort you will find in the shops.

Another rumour is that solar cells are ecologically unviable because they require such a large amount of energy to produce. I have information that, depending on the methods of production, solar cells will recoup the energy used within 3 to 7 years.

Poly and mono crystalline cells have now been in use for around 40 years and (apart from those which have been damaged) are still working perfectly. The very early solar panels had some problems with broken back covers, so water or humidity could damage the silicon cells. Plus some old series of AEG solar panels were made with aluminium connectors between the single cells inside the panel which also caused problems after 10 years or so. All modern solar panels use silver connectors and do not have these problems.

Every single silicon plate in the panel increases the voltage of the panel by 0.55V. Every solar cell has more or less this same voltage regardless of its size because the size of the solar cell affects the power in amperes but not in volts. Therefore to create enough tension (volts) to charge batteries it is necessary to connect a series of many solar cells together, with the bottom (+) of each cell connected to the top (−) of the next one. In this way all the volts are added together. For example 36 solar cells will produce 19.8V when the sun is shining on them. This is the voltage required in order for the electricity to flow into the battery, and therefore most solar panels sold are comprised of 36 cells. The solar plates are combined to make large modules behind special glass coated in PVC to protect it from the weather. The frame is made from aluminium or stainless steel. On the back are the plus and minus connections for the wires. In many solar panels there are also one or more diodes which ensure that energy continues to flow into all the solar cells even when some are in shadow. If you connect two or more solar panels together it's important to separate them with diodes (see the chapter, Connection Plans, page 59). This is useful if you connect solar panels of different output voltages, like different amount (36 and 40) of cells or anytime when there is more than a 2V difference.

The biggest natural enemies of solar panels are wind, children and bicycles, so your panels need to be very well installed and ideally high up and out of reach. PLEASE do not let them lie on the ground or lean against a wall. Fix them securely on a roof so that they are safe in storms. Make sure that they are somewhere where shadows (from trees, pylons ect.) won't fall on them because the shadow of even a few leaves or bird droppings will cause a considerable loss of power. It's good to clean your solar panels from time to time to ensure the greatest possible energy from them. In summer you may get more energy than you need so it's good to fix your panels at the correct angle for the winter sun rather than the summer sun. The panel should be at a right angle to the winter sun.

You can find this angle by putting the corner of a book at right angles (90°) to the panel and tilting the panel until the shadow of the book disappears (see picture).

For travelling people it is easier to keep the panels flat on the roof, otherwise you have to adjust the position of the panel every time you move. They are also less conspicuous this way and less likely to be stolen. And when the sun is not shining the panel gives out the most in the flat position. However a good compromise generally seems to be 45°.

In summer the sun makes the panels very hot, and cell temperatures of over 65°C may actually reduce the power by up to 75%! So it's good to have a gap of at least 3cm beneath the panel and the roof, so that the breeze will be able to take away the heat and keep the panels cool.

It can be clever to position mirrors either side of the panels to reflect the light in from different angles. This is even more effective when you construct a simple machine which will move the panels automatically

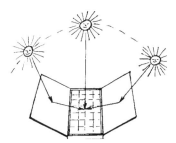

with the sun. (see also the chapter, Sun-following Systems, page 95). But for many of us this might prove difficult to build and require more energy to construct than you will gain from it.

If you make sure that your solar panel is properly fixed, in a safe place where it will not be damaged, you may have it for the rest of your life!

The cost of a 50W panel is normally around £300 but many solar associations buy panels in bulk and so it is cheaper to buy though them. A 50W panel is usually enough unless you have equipment requiring a permanent electricity supply (e.g. a fridge). For lighting, music and occasional use of other small equipment (such as a sewing machine, mixer or laptop) a 50W panel can be enough. For calculating how many panels you'll need see the chapter, Watt and Volt, page 39.

Lastly, one remark – the energy industries have bought up most of the solar panel factories in order to set high prices and dictate the time when solar energy becomes more affordable. It's not really possible to say here which factories are independent of these fossil fuel industries because this is always changing.

Batteries

You can use solar electricity direct – for instance garden fountain pumps or fans which only work when the sun is shining on the solar panel. You can also put the solar electricity directly into the National Grid but this is very expensive and you need the consent of the Grid. For this you must have a second electricity meter to measure how many kilowatt hours you put into the Grid, and the end of the year you pay the balance, if you have used more from the Grid than you have put in. This is useful to avoid the toxic waste from old batteries but the downside is that you still rely on the Grid and you still have the unknown dangers of 'electrosmog'.

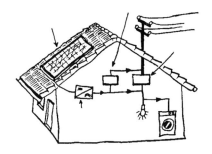

Another possibility is solar hydroelectric power because then the energy can be stored as water, which has the potential for electricity on demand. You use the sun to pump water from a low point to a high point and then release the water through a microhydo generator when you require the power. A wonderful idea but for small groups of people this is far too expensive. A quite new idea is to store the solar power with hot oil (up to 200°C) in big tanks and produce electricity when needed from the hot oil using a sterling motor. You can also cook with the oil when you let it flow inside the stove plates. See www.tamera.org for more information.

So the only viable option left to store the electricity is still good old batteries. The problems of batteries have been going on for 30 years – they have too low capacity, are too heavy, produce toxic waste and are expensive. Researches have been trying to produce better batteries (e.g. for electrical cars) but so far no alternative has been found. All types of batteries are 100% recyclable and so in theory there are no problems. However, in reality they are rarely recycled properly. One way to reduce waste is to take good batteries from scrap yards, but generally it's better to compromise and buy new ones.

If you treat your battery well it will last you for many years. There are also a couple of things you should know about the different types of batteries in order to extend the lifetime as much as possible.

Batteries should have a big capacity i.e. be capable of storing a lot of energy – capacity is measured in amp hours (Ah). In small rechargeable batteries this is measured in milliamp hours (mAh).

Batteries should be capable of being charged, used and recharged over and over again. The number of times a battery can be recharged varies a lot, but you can only find out the number of so called 'charging cycles' from the manufacturer. In some cases, e.g. car batteries, the battery has to give out a lot of electricity in a very short time, for example when you start your car. Or it may be made to be able to recharge very quickly (for example in cordless tools). But not every battery type is able to do this 'high current' input or output. Batteries should also store the energy for a very long time. This is measured by the rate at which they slowly lose their charge, which is called the 'self discharging rate'.

There is a confusing profusion of batteries for different uses. I would like to talk about the most usual types which are: car batteries, solar batteries, lead gel batteries and heavy duty batteries. They are all also called Acid Batteries. Then you have small rechargeable nickel cadmium batteries, metal hybrid batteries and lithium ion batteries.

Car Batteries

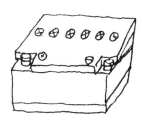

This common battery type is used to start the engine and so can give a lot of energy in a short time. Peak currents in small cars can be up to 1,000W, in trucks even up to 4,000W when you start the engine.

A car battery may be marked 12V/44Ah/175A which means that 175A is the maximum current possible; there is a capacity of 44Ah and 12V tension.

Car batteries are normally sold in black or transparent plastic cases with 6 cells, lead plates and sulphuric acid mixed with distilled water. Most of them have 6 vent caps on the top to refill distilled water. You can also buy maintenance free batteries where you will not find these refill caps on the top. But mostly they are only hidden under a plastic cover.

All the plates must be below the surface of the electrolite (diluted acid) because electricity cannot flow between uncovered plates. If the electrolite is too low the battery must be filled with distilled water and nothing else – normal water contains many different mineral ions. Only put more acid into the battery if it has spilt. But be careful – battery acid, even dilute is very corrosive, so avoid contact with the body. If you get it on any of your clothes you must wash them immediately with lots of soapy water – otherwise they will be full of holes. In general you can neutralize battery acid with soapy water. However it is a good idea to have some Baking Soda to hand to use on spills. If you intend to work with batteries regularly it is worth investing in a Hazard Kit which should include a plastic apron, rubber gloves, eye protection, acid neutralizer and cleanup materials.

You can check a battery's charge in two different ways, Firstly, you can use a voltmeter to measure the tension – 10.8V is empty and 13.8V is full. Secondly, you can use a hydrometer to check the acidity – 1.1 kg per litre (kg/l) is empty and 1.28 kg/l is full. Hydrometers are very cheap and you can buy them in most petrol stations. Don't forget to clean the hydrometer with lots of water to wash away the acid!

However voltmeters are safer and more practical.
See also the chapter, Multimeter, page 47.

Check your battery when it has not been charged or discharged for 2 hours and look in the table to get (more or less) the actual capacity state. (this table was made with a new 88Ah battery).

Capacity %	Tension V	Acid kg/l
100	12.70	1.265
90	12.58	1.249
80	12.46	1.233
70	12.36	1.218
60	12.28	1.204
50	12.20	1.190
40	12.12	1.176
30	12.04	1.162
20	11.98	1.148
10	11.94	1.134
0	11.90	1.120

- If you let a battery get so flat that it goes below 10.8V it will permanently lose some of its capacity, because the acid will begin to corrode the lead. Flat batteries must be recharged immediately. If you don't lead comes off the plates and falls down to the bottom of the battery box. In bad cases this causes short circuits between the plates in the cells and the tension of these cells quickly drops to 0V.

- In winter a flat battery can freeze and break its case so it's a good idea to keep your batteries inside in the wintertime. Don't be scared about the gases coming out of a battery. This is usually perfectly safe so long as the solar regulator is working correctly. Your battery is safe from freezing when it's:

60% charged until -30°C
40% charged until -20°C
5% charged until -10°C

- If the battery is charged to more than 13.8V the acid will begin to bubble and the oxygen and hydrogen in the water will separate, potentially causing explosions, when the gas can't get away and there is a spark e.g. from a light switch... also the battery will lose water and can dry out. It starts to get dangerous for the life of the battery when the plates are not fully covered with electrolite.

- If your car battery is not shaken for a long time then the acid and water will separate and the plates which are left only in water may get coated in sulphur and stop working, causing the battery to lose capacity. Therefore if your car batteries are fixed (e.g. in a house) it's good to shake them every 2 months. Alternatively you can charge them to 14.4V so that the bubbles mix the acid and water. Most solar regulators do this automatically (after the batteries have been discharged below 12.2V) in a controlled and safe way. If you do this yourself you must open the top vent caps and open your windows to vent the gas emissions.

- The number of charging cycles of a brand new car battery is from 40 up to 200 (depends on the quality). This means that if you use a battery until it's empty every day it will last only between 1½ and 7 months! So it's far better to use more battery capacity, that means more batteries at the same time, and never completely discharge them. In 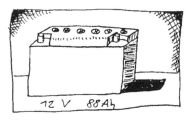 this way the number of charging cycles in a good battery will be much more than 200. It's proportionate, if you always discharge to 50% you will get 400 cycles, if you do only 20% you will get 1,000 cycles, if you do only 7% you will get 3,000 cycles.

- Car batteries can last anywhere between 3 and 9 years. Used daily in a solar system this means 1,000 to 3,000 charging cycles. So you need to have a battery capacity 10 times higher than your usual daily consumption. For more about this see also the chapter, Watt and Volt, page 39). Car batteries are the cheapest batteries you can buy because so many are produced.

- However the self discharging rate of car batteries is very high – for example a new 88Ah (amp hours) battery loses 10 milliamperes which means that if it's never charged it will be empty already after one year and remember, below 10.8V it will start self-destructing by loosing capacity due to the acid eating up the lead.

Solar & Leisure Batteries

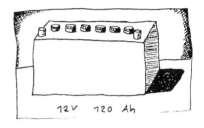

Some are specially made for use in solar systems while others are made for use in boats and caravans. They are both constructed in almost the same way as car batteries but are better because they have up to 500 charging cycles and only half the self-discharging rate. This means it will be 2 years until an unused solar battery is empty. Because they have twice the lifetime, they cause less toxic waste.

When empty (under 10.8V) they don't lose their capacity as fast as car batteries. The disadvantages are that they cost more (but not double although they last twice as long) and they have a lower maximum current (ampere) output than car batteries. One single solar battery is not capable of running machines which require a high amount of electricity (e.g. big power inverters using 1,000+ watts, or circular saws made out of car starter motors). Note the maximum output (max. ampere) written on the battery. You will have to connect two or more solar batteries together in parallel if you want to use a powerful inverter.

For this reason they can't be used as car starter batteries. The battery is simply not strong enough and the connections inside would melt if you tried. But they are highly recommended for solar systems.

Lead Gel Batteries

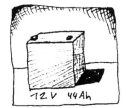

These are made for use in situations where the
battery will be in many positions (e.g. a wheelchair).
The battery acid is between the lead plates in a
gel so that it can't leak. You do not need to check
the acid at all. Lead gel batteries don't lose their capacity when they
fall below 10.8V. although they will be destroyed if they are totally
discharged (0V).

They are normally in grey boxes without screws on the top. However
the very high cost of these batteries is disproportionate to the self
discharging rate and the number of charging cycles which is only the
same as solar batteries.

Cell Temperature	Max Voltage
+40° C	13.5 V
+30° C	13.8 V
+20° C	14.1 V
+10° C	14.4 V
0° C	14.8 V
-10° C	15.3 V
-20° C	15.8 V
-30° C	16.2 V

They are very sensitive to overcharging. They mustn't be charged to
more than 13.8V (and in the summer even less than that, see also
temperature table!) because when the acid inside starts to make bubbles
they can't dissipate and stay inside insulating the gel from the plates.
The battery quickly loses capacity and in severe cases the battery box
will expand so it looks like it's been pumped up.

• If you use these types of batteries in hot places (e.g. beside a
generator, next to your stove, or in the south of Spain) the gel may
dry out and then they will not give any power – even when fully
charged.

Deep Cycle Batteries

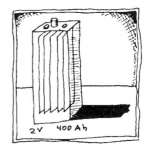

These are normally used in industrial systems (e.g. in fork lift trucks and milk floats). But they are also used more and more in professional solar systems. They have large capacities: 500 up to 2,500Ah. They are huge, heavy, robust and long lasting but also very expensive. Mainly they are made of single 2V cells in a transparent plastic box. Usually you use 6 of them connected together to get 12V (or 12 of them in a 24V system, etc. Larger solar instalations often operate at even higher voltages usually 36V or 48V).

They can be discharged without losing their capacity because the plates are usually covered with a plastic net preventing the lead from falling down. They can have up to 1,000 charging cycles so they are perfect for fixed solar systems but are mostly too big and heavy for mobile ones.

In General

Generally acid batteries will have a maximum efficiency of 80%. That means you will always put 25% more into it than you can get out again. And all batteries lose capacity during their lifetime. This means that an old battery will reach the maximum tension of 13.8V quickly after only a few hours of charging, and drops to the minimum tension of 10.8V very quickly after using only a few lights in the evening. So an old battery is quickly empty in the evening and quickly full in the day. A new battery will charge slowly and store a lot more for using later.

Often a battery starts dying by losing capacity in just one cell. The cell gets full very quickly, the tension in this cell then rises and it starts bubbling. This results in a higher tension in the whole battery (all cell-tensions added together) and the solar regulator stops the charge from the solar panel before all cells reach their maximum voltage. The battery doesn't really get full and will deep discharge more and more often, which destroys the capacity even more quickly.

Another way batteries die is sulphur built up on the plates (see also Tips 4 and 5, page 103). This happens when the battery is not used properly for a long time, or used only with very light loads. It's good for a battery to take out a big current (within the capacity of the battery of course) from time to time to crack up the sulphur built up on the cells. Or you can also use Battery Pulsers which crack the sulphur layer by short bursts of up to 80A for a few milliseconds. The 80A pulses are so short that these Pulsers use only around 100 milliamperes.

If worn out or damaged batteries are used in a solar system the solar regulator may be damaged. This is because it can't put the power from the solar panel into the batteries causing it's components to overheat. If the regulator gives up the output voltage may become too high for the connected appliances.

Check List For Second Hand Batteries

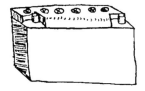

If you find a used lead battery then I recommend this checklist to determine whether it is worth using.

Quickest at the scrap yard:

1. Check the voltage – if it's under 10.8V, it's not good if you don't know how long the battery has been like this.

2. Bit dangerous but good quick check is to short circuit the poles with a small cable (e.g. $0.75mm^2$ / 0.5m long). If the cable gets hot and you see some sparks, it's good. For a better testing method see also the chapter, Battery Tester, page 100.

3. Check the acid – are all the plates covered? Is the level in all the cells more or less the same? Is the weight of the acid the same in every cell? If not, one or more cells might already be destroyed.

Test at home:

1. Charge the battery up fully – measure the voltage when you first connect the battery to the solar panel or battery charger. If the voltage rises up very quickly over 14V the battery no longer has any capacity. After being charged the weight of the acid should be 1.28kg/l and the voltage shouldn't drop below 12.6V after one hour. Comparing the battery voltage with the acid weight (see the table on page 22) gives you an idea how good the battery still is.

2. Fix the battery to a strong load (e.g. 4A = 50W lamp) the voltage should drop not more than 0.5V.

3. Then try the same with something really strong like a car starter motor (e.g. 100A = 1,200W). It's not good if the voltage drops under 9V.

4. Also watch the self discharging – if it's more than 0.1V per day it's not very good. You can check the real capacity only when you connect the fully charged battery to a one ampere lamp (12W) and count the hours until the battery is down to 10.8V. The number of hours giving out 1 ampere (A) is the capacity of the battery in amp hours (Ah).

WARNING!

When you remove a car battery from a car, first disconnect the minus (-). Otherwise you could create a heavy short circuit with the tool, from the plus (+) to the body of the car which is minus! The tool may get very hot, glowing in just a few seconds. If this happens the battery can get so hot that it explodes.

When you connect a new battery in a car go reverse, first connect the positive (+) and then the negative (-).

Nickel Cadmium (NiCd)

These are a useful replacement for small one way (non rechargeable) batteries. They are sold in the same sizes as normal 1.5V batteries.

• NiCd have a very low self discharging rate and with computer battery rechargers you can get up to 5,000 charging cycles from them.

• They can give a lot of energy in a short time so they are also used in electric cars. However they are so expensive that they are generally only used on a small scale (cordless drills, older mobile phones, etc).

• They are highly toxic because they contain a high heavy metal count of cadmium and quicksilver and must be returned to the makers for recycling.

• The voltage is 1.2V, a bit lower than the normal 1.5V batteries. This can be a problem, although generally it isn't.

• The capacity of a NiCd is only half of Zinc-Carbon batteries and a quarter of Alkaline batteries. So they have to be changed and recharged much more often than with normal one way batteries.

• NiCd do not have the problem of permanent loss of capacity because they simply stop giving out power below the deep discharging level. This means you can just use appliances until they stop working without worrying about the battery running too low. But it is important to remove the batteries from the appliance at this point otherwise they will discharge below the safety level.

Unlike lead batteries, you should use NiCd until they are empty before recharging them, otherwise you will get a 'memory effect'. It is a strange attribute of NiCd that if you always discharge then to a certain point they will 'remember' and never go beyond this point – like a stubborn donkey who is used to walking a certain distance every day and refuses to go any further. You can eliminate the 'memory effect' again with 2 or 3 full charges and discharges. With special recharging systems (reflex or CCS charging) memory effect is eliminated within one recharge.

However if NiCd are charged with simple constant electricity as in simple rechargers then you must discharge them down to 0.9V per cell. Check with voltmeter! (see the chapter, Multimeters, page 47). Most NiCd are able to be recharged quickly. When you use the quick charging method it's very important that the voltage of each cell doesn't rise above 1.52V, otherwise they will cook and get damaged inside.

Here are the normal charging times when you charge with a constant current, like the Battery Charger on page 118. You just have to look for the capacity written on the battery (e.g. 1,200mAh) and multiply it with the rates of the charge type. The standard charge is always the best.

Standard charging: 14 to 16 hours at a rate of 0.1x of the capacity.

Quicker charge: 4 to 6 hours at a rate of 0.3x or 0.4x of the capacity.

Quick charging: 1 to 1.5 hours at a rate of 1x or 1.5x of the capacity.

Hold charging: Continuous at a rate of 0.03x or 0.05x of the capacity.

With One Way 1.5V Batteries V =	Number of Cells in the Appliance	With Re-chargeable 1.2V Cells	Max. Charge Tension (V)	Discharge until (V) (only NiCd)
1.5	1	1.2	1.52	0.9
3.0	2	2.4	3.04	1.8
4.5	3	3.6	4.56	2.7
6.0	4	4.8	6.08	3.6
7.5	5	6.0	7.60	4.5
9.0	6	7.2	9.12	5.4
10.5	7	8.4	10.64	6.3
12.0	8	9.6	12.16	7.2
13.5	9	10.8	13.68	8.1
15.0	10	12.0	15.20	9.0
16.5	11	13.2	16.72	9.9
18.0	12	14.4	18.24	10.8

Nickel Metal Hybrid (NiMh)

These are the next generation after the NiCd and have 50% more capacity, no memory effect and no toxic metals but they are more expensive. They also have a higher self discharging rate and are not able to give such high amounts of energy in a short time. The voltage of each cell is 1.2V the same as NiCd. Because they are less toxic to the environment, it is a good idea to use them whenever possible. You can charge them in much the same way as the NiCd but they don't like quick charging.

When NiCd or NiMh are recharged it takes about 1.5 times the amount of energy they subsequently give out. So the efficiency is only 65%. Even so, for mobile use they are indispensable.

Tips & Tricks

• For old, used NiCd or NiMh: After a full recharge, holding a wire from plus to minus creating a short circuit just for a couple of seconds, will 'wake up' the battery and increase a number of charging cycles again.

• If for example you forget to switch off your torch and the batteries are now totally empty, you can try to rescue them by shock charging each single cell for a few seconds directly from a 12V battery. Don't confuse the polarity, don't let them get too hot, and continue charging them normally afterwards.

Lithium Ion Batteries (Li-Ion)

This type is used in mobile phones, laptops, digital cameras, MP3 players etc. The capacity is very high in comparison to the size. They can also reach 2,500 charging cycles or more before they lose their capacity.

They don't like quick charging at all. It is best only to charge them in a special charger. It is said that they have no memory effect but this doesn't always seem to be true. The Lithium inside is toxic waste, so please return them back to the store or manufacturer for disposal.

These are the most common sizes of small batteries:

Micro AAA: 11x45mm
Mignon AA: 15x51mm
Baby C: 26x60mm
Mono D: 33x62mm
9V-Block: 48.5x26.2x17mm

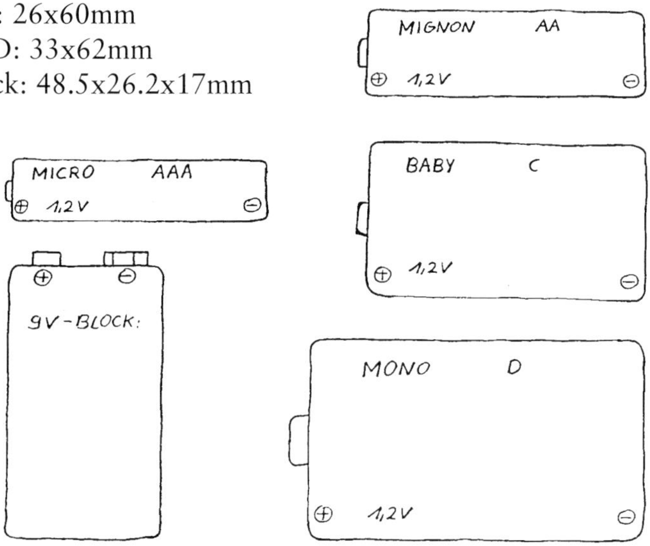

MIGNON   AA

⊕ 1,2 V   ⊖

MICRO   AAA
⊕ 1,2 V   ⊖

BABY   C

⊕ 1,2 V   ⊖

⊕   ⊖

9V - BLOCK:

MONO   D

⊕   1,2 V   ⊖

Regulators

In the last two chapters we saw that batteries are very sensitive to deep discharging and overcharging and that solar panels have a very high charging tension which could cause the battery to 'cook'.

Also at night electricity flows from the battery back to the panel causing the battery to be slightly discharged. Therefore it is better to put a diode between them so that the electricity can only flow in one direction. A special Schottky-Diode is built into every solar charger.

To avoid having to regulate the input and output by hand it's a good idea to put a charge controller between the battery and the solar panels and a deep discharging regulator between the battery and the appliances. We could do the work of these regulators ourselves, but it would involve checking the level of the battery all the time and disconnecting the panel at night. Most of us have better things to do, so regulators are more of a necessity than a luxury.

Regulators are very useful in prolonging battery life, so they are an import-ant factor in the environ-mental impact of solar power systems.

Solar regulators are high tech electronic products. Enthusiasts may wish to make their own simple solar regulators at home (see the chapter, Home-made Regulators, page 119). But for most of us, if you compare the time spent making one with the small cost of buying one, it hardly seems worthwhile. There is a compromise – you can buy regulator kits which are cheap and very usable, although they are not of the quality of micro computer regulators.

Often charge controllers and deep discharging regulators are combined in one unit. There are several different types of regulators...

Two Point Regulators

These switch off the solar panel when the battery has more than 13.8V. When the battery voltage falls somewhere below 13.2V it switches the solar panel back on again. In this way the panel alternates between on and off towards the end of charging and the battery becomes fully charged very slowly.

Many of these old regulators are made with mechanical relays, which make a noise and also have a limited lifespan, they might stop working after only a couple of years. An advantage of this type of regulator is that it switches very slowly and so only produces a very small electrosmog field.

One step better than simple 2 Point Regulators are regulators which send the excess power into a second output for another battery or an appliance such as a ventilator or heater. This is very practical in summer when there's lots of excess power.

Shunt Regulators

Are the most common regulators today. Also known as an U/I Reference Line regulator. It is slightly more sophisticated than the two point regulator and regulates the power from the solar panel so that the voltage of the battery is always maximized. The battery receives exactly the amount of power it needs in order to reach 13.8V. These regulators charge batteries to the top as fast as possible.

The Shunt Regulator creates a short circuit in small pulses to the panel so that excess power only flows between the regulator and the panel and not into the battery. This pulsing with frequencies of 50 to 2,000 Hertz causes large electrosmog fields around the wiring between the solar panel and the battery. This can disturb phones, radios, computers, etc. (see also the chapter, Electrosmog, page 73).

Gas Control Regulators

If lead batteries are not shaken from time to time (such as when you drive your car) the slightly heavier battery acid will slowly separate from the distilled water. So the acid sinks down and the water goes up. Regulators with gas controllers increase the maximum battery voltage to 14.4V or even 14.8V usually after being discharged under 12.3V or 12.4V, this causes the acid to bubble and mix with the distilled water again.

Attention!! If you use lead gel batteries with one of these regulators you must switch off this gas control mechanism because the maximum voltage of lead gel batteries is only 13.8V!

Micro Computer Regulators

These are the best for your batteries because they can adapt to the actual temperature, capacity, age and load level of the battery and therefore maximise the life of your batteries.

Micro computer regulators sometimes have a couple of digital measuring instruments inside like a voltmeter for the battery, an ampere meter for the current coming from the solar panel, another ampere meter for the current going out to the appliances, and amp hour meters for the total solar income and the total battery output.

But these meters also have a downside. For example: if you use appliances directly from the battery like 230V inverters or solar pumps, the built in amp hour or capacity meter might get totally confused and won't show you the correct load level of your battery any more, because it measures and calculates only the current flowing directly through the regulator. Despite this they usually do a good regulating job and the voltmeter will still work correctly.

In General

The maximum solar panel power of a regulator (e.g. 10A) will be printed somewhere on the regulator. If you use it for a higher power the internal fuse should blow. If it doesn't the circuit board or the Mosfet-Transistors in the regulator will get too hot and can burn out instantly. So never put in a bigger fuse than allowed.

Panel-Power In W (watt)	Maximum Current in A (ampere)
50	2.50
75	3.75
100	5.00
150	7.50
200	10.00

Never confuse the polarity of the solar panel or the battery, most regulators can't stand this treatment at all.

You can use two or more regulators on the same battery if one regulator is not strong enough for all the solar panels you have.

If you use more appliance load than the regulator is rated for, you will blow the internal fuse. So before choosing a regulator calculate all your appliances together and look in the following table to see what ampere rating is required. (this table is only for 12V systems).

Appliance power	Max. ampere
75W	6.3A
96W	8 A
120W	10A
180W	15A
240W	20A
300W	25A
360W	30A

Deep Discharging Regulators

Most solar regulators include a deep discharging regulator which switches off the appliance, e.g. lamps, when the battery reaches its deep discharging level of 10.8 or 11V. They switch on again when the battery reaches around 12.5V or on some old style units only after reaching 13.6V.

Deep discharging regulators have a maximum output usually between 4A and 30A (= 50W to 360W). This is OK for lamps and small music systems but not for much more. For more powerful appliances, e.g. a battery powered drill, it is best to use a relay (see the chapter, Connection Plans, page 59). But if you wish to use very powerful appliances, such as circular saws made from car starter motors which use up to 250A, you will need to connect them directly to the battery. Use small appliances such as a lamps as indicators of when to stop using the directly connected load!

To make a simple deep discharging regulator see the chapter, Home-made Regulators, page 119.

Most 230V inverters have a built-in deep discharging regulator which works totally independently from your solar regulator.

When you have a very old solar regulator (e.g. with relays) it's a good idea to check how much electricity the regulator itself requires. This often isn't in the instructions and you will have to check with a multimeter (see also the chapter, Multimeters, page 47).

Most modern ones require only 2-15 milliamperes while old ones may well take 100 milliamperes or more.

For example the deep discharging regulators you get for small car refrigerators are very cheap but use a lot of electricity because they use a relay (electromagnetic switch).

Bi-polar relays are much better because they work with only a short electrical impulse.

Modern solar regulators use so called 'Mosfet' transistors which consume hardly any power.

The Most Common Mistakes

1. Never confuse plus (+) and minus (−) on regulators, not only the fuse but also important components could be destroyed.

2. Never leave the regulator connected to the solar panel when the battery is disconnected. The regulator would get all the high tension of the solar panel, not be able to get rid of it and overheat.

3. Electronics are very sensitive to moisture so it's good to have a well sealed box around the regulator.

4. Use short, thick cables to connect the battery to the regulator, and use good...

5. ...Pole clamps with a bit of fat or Vaseline. This avoids the loss of tension created by the resistance of the pole clamps together with the battery poles. If you don't do this the regulator will not be able to measure the battery voltage correctly and will tend to switch the panel off too early or give up working completely.

6. If you create a short circuit or confuse the poles, the regulator fuse should blow – so always check the fuse before deciding your regulator is broken.

Watt & Volt

An electric current is composed of electrons and is totally natural. Clouds produce electric charges which discharge in thunder storms; our nervous systems use electricity to convey messages of movement to different parts of the body. However most people are not familiar with the basics of electricity. Even those of us who learned them at school have generally forgotten them – so in this chapter we will look at the basics of electricity for solar power systems.

It is important to be able to calculate the size and efficiency of a solar system and to do this we must dive into the strange world of volts, amperes, watts, ohms, electric power, kilowatt hours, and parallel and series connections along with all their equations...

I am not going to be using the normal scientific symbols because they are too complicated – instead I am using V to stand for volts, and so on.

You cannot see or hold electric energy itself, only its effects and so many people are curious about exactly what it is. We can more easily understand electric units when they are compared to something we can see – here I use a model of a hydroelectric power station to explain the law of Ohm, electric energy and power.

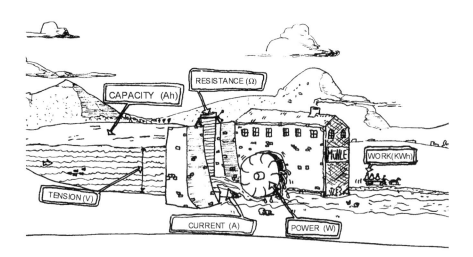

Electric Power

The *level* of the water is the electrical tension which is measured in *Volts (V)* and the *volume* of the water pouring through the sluice gate is the electrical current which is measured in *amperes (A)*. The combination of *water level (tension)* and amount of *water (volume)* produces the electrical power which is measured in *watts (W)*.

When the water level is higher (more tension) there is more force behind the water stream (like higher voltage). The volume of water (amperes) pouring through the sluice gate also affects the power of the turbine (watts). This can be written as an equation:

$$\text{tension x current} = \text{power}$$
$$\text{volt x ampere} = \text{watt}$$
$$V \times A = W$$

This could also be written as:

$$\text{Volt} = \frac{\text{watt}}{\text{ampere}} \quad \text{or} \quad \text{Ampere} = \frac{\text{watt}}{\text{volt}}$$

The Law Of Ohm

When the water level is higher the pressure on the sluice gate is higher and the gate doesn't need to be opened very wide in order to run the turbine. On the other hand, when the water level is lower the gate needs to be opened wider for the turbine to run with the same power. So the opening of the sluice gate can be compared to electric resistance which is measured in Ohms (Ω). The smaller the opening, the higher the resistance. The volume of water (A) coming through the sluice gate is the product of the level of water (V) combined with the diameter of the sluice gate (Ω).

So we have the following equations:

$$\text{Current} = \frac{\text{tension}}{\text{resistance}}$$

$$\text{Ampere} = \frac{\text{volt}}{\text{ohm}}$$

$$A = \frac{V}{\Omega}$$

Also the calculated equations:

$$\text{Volt} = \text{ampere x ohm}$$

and

$$\text{Ohm} = \frac{\text{volt}}{\text{ampere}}$$

Both of these basic equations can be remembered with the magic triangles. Cover the wanted outcome with one finger and the necessary calculation will be visible.

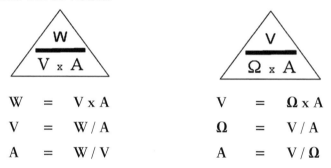

W	=	V x A
V	=	W / A
A	=	W / V

V	=	Ω x A
Ω	=	V / A
A	=	V / Ω

Volt and ampere are connected to watt and therefore both these magic triangles can be combined to produce further equations. Here is a table hammered in stone of all the variations:

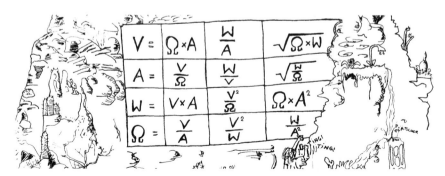

41

Some Practical Examples

1. To work out how many watts an electric motor uses which pulls 10A from a 12V battery just calculate 10 x 12 for the solution 120W.

2. A solar panel giving 20V with a current of 2.5A will be 50W. (2.5 x 20 = 50)

3. A 20W lamp with a 12V battery needs a current of 1.66A. (20 / 12 = 1.66)

4. A 5W Radio on a 12V battery needs how many amperes? A = W/V ... 5W / 12V = 0.416A (which can also be written as 416 milliampere – mA)

5. A 12V soldering iron with 30W power has a resistance of...? $V^2 / W = \Omega$ ... $12^2 / 30 = 144 / 30 = 4.8\Omega$
and a current of...?
A = W / V ... 30 / 12 = 2.5A
A = V / Ω ... 12 / 4.8 = 2.5A

Capacity

To find out how long an appliance will run until the battery is empty you must know the power (W) in addition to the time (in hours). The capacity of a battery is shown in amp hours (Ah). On small batteries it is shown in milliamp hours (mAh). A truck battery with 100Ah capacity will run an appliance for 100h with 1A or 1 hour with 100A.

1. A 20W lamp pulls 1.66A from a 12V battery so a 100Ah battery can run it for...?
Ah / A = hours ... 100 / 1.66 = 60 hours

2. An electric motor using 10A will run for...?
Ah / A = h ... 100 / 10 = 10 hours

3. A solar panel charged for 10 hours on a sunny day with 2.5A will give...?
hours x A ... 10 x 2.5 = 25Ah per day

So theoretically it takes 4 days to charge a 100Ah battery with a 50W panel. In reality it takes 25% longer than this. So it is 5 days, because even a new battery is only 80% efficient. From this 100Ah input it will only store 80Ah. So you have to put in 125Ah to keep 100Ah for use.

The older the battery, the less efficient it will be, but anyway we always have to put more energy in than we will get back out.

Kilowatt Hours (kWh)

You might recognize kilowatt hours (kWh) from electrical bills. It is calculated by multiplying power (W) with time i.e. watts or a thousand watts (kW) multiplied by hours. If an appliance with 1,000W runs for an hour it requires 1kWh. It is possible to calculate a battery's potential (kWh) by multiplying the capacity with the voltage.
e.g. 100Ah x 12V = 1,200Wh or 1.2 kWh

Connections

You can connect solar panels, batteries, loudspeakers and appliances in many ways. But unless you know what you're doing you may cause a short circuit. There are two main connection types parallel connection (side by side) and consecutive connection also called series connection (one after the other).

Parallel Connections

For instance, to enlarge your battery capacity you could connect two batteries together in parallel. It's best to connect only batteries of the same size and age, because otherwise one may discharge the other. When you connect a small and a big battery or a very old battery with

a new one the old one probably has much less capacity, even th
originally they were the same Ah. Electricity will always flow betw
them when you charge or discharge them in order to compensate
the different capacities and because batteries are only 80% efficie
20% of this electricity will be lost each time.

Total = 12V 300Ah

Three 12V 100Ah batteries connected in parallel.

In parallel connections the capacity of each battery is added together
but the voltage remains the same.

Normally appliances would also be connected in parallel.

Three 12V appliances connected in parallel.

The power (W) and the current (A) used by appliances connected in
parallel is added together.

So in our example: 5W + 20W + 20W = 45W
                    416mA + 1.6A + 1.6A = 3.736A

44

ᵒᵘᵍₕ
'ₑₑₙ
fₒᵣ
ₗₜ,

nains the same. But the total resistance gets smaller
quation:

$$\text{Total } \Omega = \cfrac{1}{\cfrac{1}{\Omega 1} + \cfrac{1}{\Omega 2} + \cfrac{1}{\Omega 3}}$$

$$\text{Total } \Omega = \cfrac{1}{\cfrac{1}{28\Omega} + \cfrac{1}{7.2\Omega} + \cfrac{1}{7.2\Omega}} = 3.19\Omega$$

. there are only two appliances in parallel and they have the same resistance, it's much simpler to calculate. If you take the radio out of our calculation then the lamps' resistance is simply halved:

$$\frac{7.2\Omega}{2} = 3.6\Omega$$

Series Connections

Are totally different. For example, you can connect small 1.2V rechargeable batteries consecutively to produce a higher voltage.

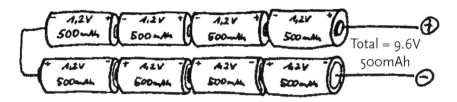

Eight 1.2V 500mAh batteries connected in series.

In this connection all the tensions (V) are added together, while the capacity (mAh) remains the same.

When appliances are connected consecutively the resistances are added together, so less current can flow. So you could make one 12V lamp out of two 6V bicycle lights of the same power.

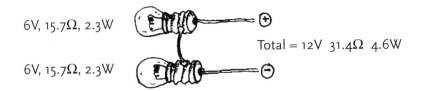

6V, 15.7Ω, 2.3W

6V, 15.7Ω, 2.3W

Total = 12V 31.4Ω 4.6W

Two 6V bulbs connected together so they can be used in a 12V circuit.

The tension is divided between the lamps in proportion to the resistance of each lamp. In this example each lamp has the same resistance and so the volts are divided equally between them.

Multimeters

A multimeter should be in the toolbox of every electronic worker, as should a soldering iron, pliers and a screwdriver. Some measuring instruments have an analog display, while others are digital, and some very simple ones only have light emitting diodes. Single meters only measure volts or amperes whereas multimeters have a greater range. Simple digital multimeters are reasonably cheap. In the middle of each multimeter is a dial which is used to set the range e.g.:

- Alternating tension (AC-volt ≈)
- Direct tension (DC-volt ⚍)
- Alternating current (AC-ampere ≈)
- Direct current (DC-ampere ⚍)
- Resistance (ohm Ω and also kilo-ohm kΩ)
- Diode test (➤⊢)

They have also 3 or 4 sockets for the measuring cables. The black cable always goes in the COM, or minus socket and the red cable goes in the correct socket for the range you are measuring.

Measuring Volts

Put the red cable in the volt/ohm/➤⊢ socket. Switch to the right type of electricity (AC for wall socket, DC for Batteries) and the right measuring range. For example, if you're measuring a 230V inverter switch to AC 750 or 1,000V, put the measuring cable in the wall socket of the inverter and the multimeter will tell you the number of volts. It differs with the type of inverter from 175V up to 245V.

If you want to measure a battery voltage switch the multimeter to DC with the range of 20V – put the red cable on the plus (+) pole of the battery and the black cable on the minus (–) pole. The multimeter shows for example 12.84V.

If you put the cable on the opposite poles the display will be minus for example -12.84V. You can check the polarity of batteries and live cables in this way. If you have an old analog instrument the needle will go in the wrong direction when the poles are mixed up.

You can check 1.5V non rechargeable batteries or 1.2V rechargeable ones on the 2V or 2000mV range. If the voltage shown is less than 1V the battery is empty.

When the voltage exceeds the measuring range (e.g. if you measure a 12V battery in the 2V range) the digital multimeter will simply show '1' on the display. However the needle of an analog multimeter may go so fast towards maximum that the needle can be damaged.

Measuring Ampere

This is very different from measuring volts because it's necessary to create an electric circuit, e.g. a lamp which is switched on, a solar panel with the sun shining on it, or an electric motor which is running.

Open the circuit, e.g. by switching off the lamp, removing a fuse, or disconnecting a cable from the battery, and then close the circuit again over the multimeter, so that all the power is flowing through the multimeter. To do this plug the red measuring cable into the 10A socket in the multimeter, and switch the dial to the 10A range.

ATTENTION! There is a direct connection between the COM and the ampere sockets, so if you leave the red cable in the 10A socket and then later by measuring volts, e.g. on a battery, you will blow it up or burn out the measuring cables. So don't forget to always put the red cable back into the volt/ohm socket when you are finished.

If you don't know the size of the current, begin always on the 10A range and then move down to mA ranges. Usually the 10A socket is not fused, but the mA socket is fused with e.g. 2A. So when you leave the range on 2000mA and put the cables on a battery the internal fuse will blow.

When this fuse is blown, you can't measure mA. No current can flow through the multimeter and the display will only show 0.00mA.

Measuring Ohms

To measure resistance the multimeter puts some volts from its internal battery in and measures how much it gets back. From the difference it calculates the resistance. Therefore it is very important that there is no tension in the circuit itself as this will confuse or in some cases destroy the multimeter. It's best to remove the battery, solar panel, main-fuse etc. before testing the resistance of a circuit.

The 200Ω range is used most often. It's used for cables, fuses, switches, motors, soldering irons, and so on... Before measuring in this range put the ends of both measuring cables together. A digital multimeter will display around 0.03Ω (that is the resistance of the measuring cables). On analog instruments there is a small wheel you need to turn until the needle is on zero.

The resistance of an electrical conductor varies according to the temperature. It goes up when the temperature is higher, e.g. the resistance of a filament in a lamp is about 10 times higher when it's switched on and glowing with 2,000°C, than when it's cold.

For loudspeakers, you can measure the resistance with the small direct current (DC) from the little battery of the multimeter, but the Ohms written on the loudspeaker refer to alternating current (AC) which is called the impedance, which is about 30% higher than the DC resistance.

Testing Diodes
(read this only if you now need to repair things!)

When the dial is switched to diode test ➤⊢ the multimeter will show '1' on the display. Put the red cable in the volt/ohm socket and connect the two measuring cables with either end of the diode. Diodes are very easy to recognise once you know what they look like. They are components in an electrical circuit which only allow the electricity to flow in one direction. The most common diodes are silicon diodes which have an average loss of tension of 0.5V to 0.8V. Special Schottky-

Diodes only have a loss of tension of between 0.1V and 0.2V and therefore they don't get so hot. The maximum current which can flow through a diode can only be found by consulting specialist tables. However it is possible to estimate the maximum current by looking at the dimensions of the diode and the thickness of the wires coming out.

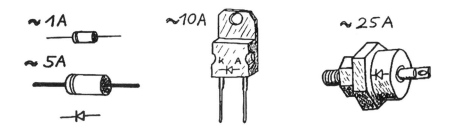

A selection of types of diode with the markings which indicate the direction of current flow. In this case they are all marked that the flow is from right (+)A anode to left (–) K cathode.

The connecting sign of a diode is ➤⊢. It shows the direction of the electricity flow from plus to minus. On small diodes the direction is shown by a small ring. Sometimes there is a sign of A and K which means anode and cathode.

(+)A ➤⊢ K(–)

The multimeter runs electricity from plus to minus through the diode. If the red cable is held on the anode and the black cable on the cathode (ring side), electricity can flow and the multimeter will show the loss of tension in mV (silicon diodes 500-800mV; Schottky-Diodes 100-200mV). When the cables are switched around no electricity will be able to flow because the diode blocks it, and the display remains on '1'.

If the multimeter displays '1' in both directions, or there is a loss of tension in both directions, or even worse no loss of tension at all in either direction (the multimeter shows '.000', that means short circuit) then the diode is dead and has to be replaced.

Testing Transistors

Transistors and Mosfets are like electronic switches. As in this model, a small current from the **B**asis allows a big current to flow from the **C**ollector to the **E**mitter. But they don't switch only ON/OFF like a relays. If you change the voltage at the Basis (B) see what happens:

We have Positive (PNP) and Negative (NPN) Types. The symbols are like this:

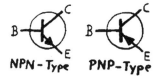

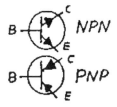

In an NPN-Transistor you can measure with your multimeter the two diodes pointing away from the Basis and in a PNP-Transistor the two diodes pointing towards the Basis. If you measure other connections the transistor will be ruined. So you can find out where the Basis is, but to find out which one the Collector and the Emitter is, you need to use look up tables.

Near to 0V the transistor blocks and no current can flow. This pre-resistance is very important. From (+) to (B) the flow is limited to a few milliamperes, otherwise there will be a short circuit between (B) and (E)

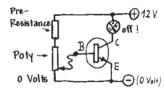

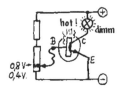

and the transistor will blow up! The critical range on normal silicon Transistors is between 0.4V and 0.8V the Transistor starts opening and a small current can flow. But the resistance in the Transistor is very high and it gets warm or even hot.

51

A higher voltage at the Basis switches the gate full on. The Transistor only has a 0.7V loss of tension, similar to that of diodes.

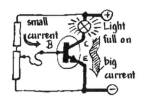

Avoid the critical range when you switch big currents, the transistor will get too hot and can be damaged. You can use two or more transistors behind each other to get a clearer signal.

Mosfets

Work quite similarly to transistors, though the connections have different names. There are positive (P-Channel) and negative (N-Channel) types, but the most common are the negative types. The Basis is called Gate, the Collector is called Drain, and the Emitter is called Source.

<u>B</u>asis = <u>G</u>ate

<u>C</u>ollector = <u>D</u>rain

<u>E</u>mitter = <u>S</u>ource

The critical range is with 3V to 7V – much higher. To run them safely use 0V to max. 2V at the Gate for OFF and 8V to 12V for ON.

Mosfets are used more and more because they only have a very small resistance when they are switched ON (about 0.05Ω, more or less like a cable!) and a loss of 0.7V like the Transistors. As they don't get so hot and don't need so much cooling, you can use much smaller heat sinks on them.

Mosfets are very sensitive – only some μA (1000μA =1mA) into the Gate switch them ON. You can use a very big pre-resistance in front of the Gate, e.g. 1MΩ (1,000,000Ω).

To test N-Channel Mosfets use the 4 following steps:

1. Switch your multimeter on at the diode test range. Red (+) measuring cable to the Source (S), black cable to the middle (Drain = D). You will measure a loss of tension between 400mV and 700mV.

2. Put the red cable to the Gate while keeping the black one at the Drain. The tension out of the multimeter now switches the Mosfet ON. The multimeter stays at '1' because no current supposed to flow here.

3. Put the red cable back to the Source. The loss of tension should be much less now, usually it is around 20 to 180mV.

4. To switch them back to OFF you do the same as in point 2, but reverse the polarity. The red (Plus) to the (middle) Drain and the black (Minus) to the Gate. After that check as in step 1. to see if the Mosfet has really switched back to OFF.

Cables

To connect a solar system up you need cabling and there are many factors to take into consideration. Solid copper cables break easily and should only be used where the cables can be fixed onto a wall. Multi strand copper cables are the best and are very flexible. Cables used outdoors, for the solar panels etc., should be of a high quality UV-resistant type.

In a 12V system there are very high currents (A) and a lot of electricity is lost when the cables are too thin or too long. Anything between a 1% and 5% loss is OK. But not more than this, especially at places where current flows a lot e.g. between the solar panel and the regulator, or between the regulator and the battery.

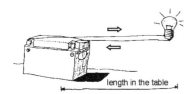

length in the table

The table below shows the most common cable diameters (in mm) and cross section (in mm²) along with the lengths which produce a 1% and 5% loss of electricity.

The two way flow to and from the battery has already been taken into account in the calculations.

Cross Sec- tion	Dia- meter	Radio 5W 416mA		Lamp 20W 1.66A		Solar panel 50W 2,5A		Motor 120W 10A	
mm²	mm	1%	5%	1%	5%	1%	5%	1%	5%
0.75	1	6m	30m	1,5m	7,6m	1m	5m	25cm	1.3m
1.5	1.5	12m	60m	3m	15m	2m	10m	50cm	2.5m
2.5	2	20m	100m	5m	25m	3m	17m	85cm	4.2m
4	2.5	32m	160m	8m	40m	5m	27m	1.3m	6.7m
6	3	48m	240m	12m	60m	8m	40m	2.0m	10m
10	4	80m	400m	20m	100m	13m	67m	3.3m	17m

The normal cross sections of cables in small solar power systems are 2.5mm² and 4mm².

Fuses

You need fuses to secure your solar system from unintentional short circuits which could easily cause a fire. Without a fuse batteries can give out such a high amount of current that cables start glowing in seconds and may well catch fire. So firstly secure all your cables with fuses. Every cable is only allowed to get electricity from the battery over a fuse which blows before the cable can get even warm. You also need fuses to secure appliances, e.g. for overload protection. Fuses should be placed as near to the battery as possible.

Use a big main fuse directly at the battery (like built in fuses in solar regulators) to secure the whole system. It has to be big enough that you can run every appliance you have at the same time, e.g. 30A (= 360W in a 12V system). But as 30A is too much for small cables like 2x 0.75mm², you need smaller fuses placed in the lines before them (e.g. 5A or 10A). There are several different types of fuses; the most common are:

The flashtube types with the dimensions of Ø 5 x 20mm. You can get them from 50mA to 10A.

The slightly bigger US-flashtube type with Ø 6.3  x 32mm is available from 200mA to 30A.

In old cars you sometimes find this type of small  ceramic bodied fuses.

And in newer cars the common flat ones which probably the best to use in our solar systems. They range in size from 1A to 40A. If you need more than 40A use two or more in parallel connection, eg: 50A fuse, 2 x 25A, or for a 150A fuse, 5 x 30A.

Fuse holders sometimes give problems with bad connections. It's best to always keep them clean and in a dry place. *Dangerous but helpful*: If you have to improvise repairing a blown fuse it's always better to solder a very small cable on to it than to wrap it in aluminium foil.

Plugs & Polarity

There is no uniform 12V plug system as in 230V systems. In 230V systems the polarity doesn't matter, because it alternates anyway 50 times per second (50 Hertz), see also the chapter, Electrosmog, page 73. But in a 12V system with direct current the polarity matters a lot.

Polarity-safe plugs are a very important safeguard against inadvertently confusing the poles of an electronic appliance, which is, in my experience, the most common mistake in solar systems. Many radios, CD and tape-players, TVs, cordless drills, amplifiers, inverters etc., have all been destroyed in this way!

It would have been easy to prevent the destruction of many of these appliances due to cross connection by putting a small, cheap diode in front of the circuits and you can easily do this. See the chapter, Music Systems, page 84.

There are some exceptions: halogen lamps and soldering irons will run in both directions but 12V fluorescent light tubes need to be wired-up correctly.

The first step to prevent confusing the poles is to use different cable colours, like black for minus (–) and red for plus (+). If the cables are blue and brown use brown for plus (+) and blue for minus (–)... like brown for red (+). and blue for black (–).

To mark cables in emergency cases make a knot in the plus (+) cable. Anyhow if you mark only one cable, then always mark the plus (+) and not the minus (–).

Great care needs to be taken that the same type of plugs are not used for 12V DC and 110/230V AC systems because you can mix them up them too easily which could prove lethal both to the operator and the appliance.

On 12V appliances use plugs where the polarity can't be confused because they only fit in the sockets one way round.

For example: cigarette lighter plugs (max. 10A).

 Or loudspeaker plugs (max. 2A)

Small jack plugs (max. 2A) work fine but only when you know which way they are connected.

  XLR plugs (max. 5A) from mixers and microphones are very good too, because you can get many kinds of plugs and sockets, like 1. Wallsocket, 2. Cable plug, 3. Wallplug, 4. Cablesocket.

If you live somewhere other than the UK you can use the normal UK three flat pin plugs and sockets (max.13A), but make sure you mark them clearly so no one uses them somewhere else for 230V AC by mistake.

There are also similar AC plugs and sockets but with three round pins (max.15A) which are usually used for theatre lighting. These are ideal for larger loads and are less likely to be confused with mains applications unless you live in a very old house or a theatre.

 In Spain and Switzerland they use a three round pin type for AC which works fine for 12V systems (max.10A). But make sure you mark these clearly as 12V plugs and use a different type of plug for AC.

Also this international IEC three pin type (max.10A) is ok, though wall sockets are hard to come by and it can be confused with AC mains systems as it is used on computers, kettles and other mains appliances.

Not OK are two pin or one pin plugs because you can easily confuse the poles, even when they are marked with different colours!

Switches

Switches are used in many different ways and there are a lot of different types, push-buttons, turn switches, slide switches, toggle switches, change-over switches, on/off switches...

Push buttons open or close the contact when you press the button. They don't stay in the position you have pressed them, they only switch on or off when you press them, like a bell-push.

On/off switches stay in the position you have switched them. You have single-or multi-polar on/off or change-over switches.

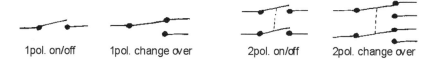

Also there are multistage turn switches with one or more poles.

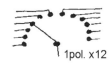

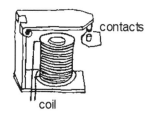

You also have electrical switched switches, they are called 'relays'. If you let a small current flow through the coil, a magnet field draws the switch and lets a big current flow between the relay contacts. You can also get relays with change over switches or more poles.

In general switches do not have an endless lifetime. 3,000 to 15,000 switchings are normal. You have to look up the max. current for the contacts, otherwise they will be destroyed much quicker. When the switched current is 5 to 10 times more than allowed they can melt!

Connection Plans

To connect our solar system you need to use all that you have learned in the last chapters about cables, fuses, switches and plugs. It's good to make the cables as short as possible to minimize electricity loss. In particular it's very important that there is not too much electricity lost between the regulator and the battery. If too much electricity is lost there will be a loss of tension, causing the regulator to switch the solar panel off too early.

If there is a high current to an appliance then a deep discharging regulator will also switch the appliance off early. For this reason good contacts on the battery poles are very important. So don't just wind the wires around the poles or use clothes pegs, or not even crocodile clamps, e.g. from jump leads. Instead use clean high quality pole clamps (soak them in boiling water to clean them if they are second hand) and fit them properly with a bit of fat or Vaseline to prevent oxidation.

If you place the regulator away from the battery (best not more than 1.5m) a powerful fuse, e.g. 30A, connected as close as possible to the battery will save the cables between the battery and the regulator in the event of a short circuit. Without one these wires could become like bomb fuses and cause a fire.

Because of contact problems in fuse holders it's even better to place the regulator so near to the battery (max. 50cm) that the cables can't connect together or scratch on anything, then you can connect without this main-fuse. The solar panel doesn't really need a fuse because it can't give out more current than it can produce. But a second fuse between the appliances and the regulator is necessary unless there is already one built into the regulator. If you want to charge different batteries it's useful to have a switch between the solar panel and the regulator so that the solar panel can be switched off when you change the battery to prevent the regulator from overheating during disconnection.

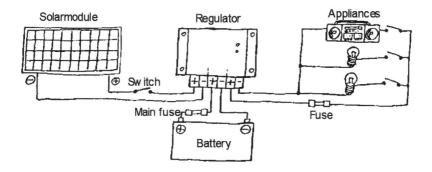

This is a basic connecting plan for a solar power system. You can build on this basic plan by adding more batteries or solar panels.

Connecting Solar Panels

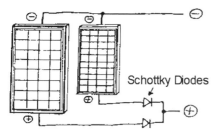

Never confuse the poles when you connect the solar panel to the regulator, if you do you will destroy the regulator. If you connect the solar panel directly to the batteries without a regulator and you confuse the poles, the diodes inside the connecter box of the solar panel will melt. In bad cases, when these diodes melt a short circuit is created destroying the lines between the single solar cells and ruining the panel!

If you want to run two solar panels with the same regulator, the regulator must be capable of taking the load. If you connect two similar solar panels that's no problem but if you want to connect different panels together you need to count the amount of cells or measure the output voltage beforehand.

If it differs by more than 4 cells or 2.5V then use two Schottky-Diodes in front of each solar panel, otherwise the solar panel with the higher voltage will try to give electricity in the one with lower tension. Put Schottky Diodes into the circuit like this so that electricity can flow out, but not into the solar panel.

Some Constructions

Home made wooden boxes are quite practical (and very German by the way). Put the battery inside the box and fix the regulator, fuses, sockets and maybe volt and ampere meters into the lid. This makes the cable from the battery 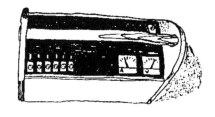 to the regulator optimally short. This is a compact, portable system of up to 88Ah (approx. 25kg) which can be used anywhere. There is an extra socket in the lid for connecting the solar panel. Using batteries in separate boxes connected with short, big cables with plugs can very simply increase the capacity.

For stationary systems it's good to make a nice protective box at eye level, containing all the electronics. You can install a big battery block nearby to provide a permanent electricity supply.

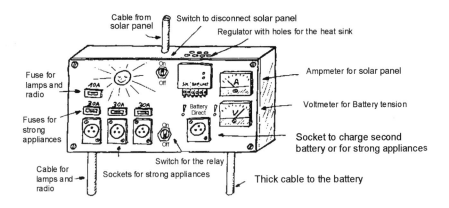

It's very convenient to also have a small battery in a lightweight box connected to the main battery block which can then be used for mobile purposes. The ampere and volt meter is a real luxury which enables you to check the actual battery voltage and the actual current from the panel. They are

Above:
12V Lamps
Made out of wood, fence wire and shells.

Right:
Walkman
With an elaborately framed mini solar panel.

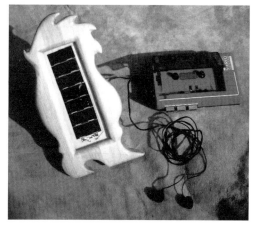

Below:
Mini Marshall Amp
With a small solar panel.

Above:
Sander
Using a car heater fan motor.

Above:
Angle Grinder
With motorbike starter motor.

Right:
Milling Machine
Made with a motor from a cordless drill.

Below:
Hammer Drill
With a powerful 12V motor connected to the original cog, cut off from the 230V motor.

Above:
Solar Grinder
Using a windscreen wiper motor.

Below:
Homemade Turning Lathe
Motor from wheel chair.

Above:
Stationary Drill
Powered by a wheel chair motor.

Above:
Solar Fan
With a windscreen wiper motor.

Above right:
Special Voltmeter
Made from an old VU meter.

Below:
Portable Solar Power Box
For totally independent use. Battery, timer, voltmeter, ampmeter, fuse-box, deep discharge regulator for output sockets, solar regulator with input socket for a solar panel.

Above:
Solar Sewing Machine
12V motor from cordless drill.

Below:
12V Solar Wash Center
With self circulating hot water collector. Old washing machine with windscreen wiper motor and relay timer for left and right spin.

Above:
Spin Dryer
Using a strong car heater fan motor.

Right:
Solar Powered Electric Guitar
With built in loudspeaker and Mini
Marshall amplifier powered directly
by solar cells glued on guitar body,
or from rechargeable batteries.

Below:
600W PA Amplifier
Power Mosfet car amp with different pre-amps and input sockets
for microphones.

Below:
600W Guitar Amp
Made with power Mosfet car amplifier and homemade pre-amp.

also useful in mobile systems particularly when adjusting the position of the panels to catch the sun.

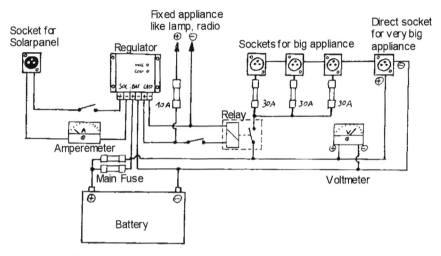

Luxury version connection plan.

In the luxury version shown above the amp meter is connected to one wire from the solar panel to measure the current, and the volt meter is connected to plus and minus to measure the battery voltage.

The deep discharging regulator switches the relay which then gives direct battery power to the large appliance. When the battery level gets too low the deep discharging regulator switches the relay off, which also switches off the direct connection between the battery and the powerful appliance. Using a manual switch between the relay and the deep discharging regulator you can disconnect the relay from the circuit when you are not using a powerful appliance – this avoids wasting the small amount of electricity used by the relay itself. A special socket with a direct connection to the battery is used for very powerful appliances (more than 30A) and for charging a portable second battery. To get a bigger main fuse you can put two or more fuses in parallel connection, so from two similar 30A fuses you get a 60A fuse.

For other constructions and connection plans for workshops see the chapter, Solar Welding, page 97.

Inverters

Inverters transform a 12V DC (direct current) into 230V AC (alternating current) which is the normal current from the National Grid. With these inverters you can run conventional appliances from your 12V system. It is often much easier to use an inverter than to change an appliance to run directly from 12V. Inverters use between 5W and 25W when switched on without any load and when in stand by mode. They can get very warm when working hard, so they need cooling, either with a built in heat sink or a ventilator. They have a efficiency rate of 70% to 90% (when they say 95%, it's mostly not true), so be aware that you always have quite a loss of electricity.

The alternating 230V current changes polarity with a frequency of 50 times per second (50 Hertz). The course of this tension follows a sine wave which looks like this:

Alternating current has the advantage of transforming easily to small and high tensions. With transformers you can change the relation of tension and current easily, like the relation between speed and power in a gear box. That's why alternating electricity is also used in electricity pylons because when you want to transport electricity over large distances there is much less electricity loss when you transform it into a high voltage (up to 300,000V) for transmission, and then transform it back to a lower voltage (230V) for use.

All inverters contain a transformer. Transformers have two different coiled cables around a common iron core. The electricity runs through the first coil which transforms it into a strong magnetic field. The iron centre transports the magnetic field to the second coil which trans-forms it back into electricity. The input and output voltage is in direct proportion to the number of turns of cable in each coil.

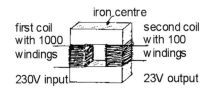

first coil with 1000 windings
iron centre
second coil with 100 windings
230V input
23V output

For example: if the first cable has 1,000 turns and 230V input tension, then if the second cable has 100 turns it will have an output voltage of 23V.

There are many different qualities of inverters:

Rectangular Wave Inverters

These are the cheapest and simplest and so also the most common inverters. They create a rectangular alternating current by a simple switching between polarities.

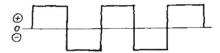

This is good enough for appliances without built-in transformers (e.g. lamps, heaters, soldering irons, coffee machines, some electric motors...). Appliances with built-in transformers are sometimes only 50% efficient when used with rectangular shaped waves. So when they are used on a rectangular wave inverter they may not work properly. The sharp corners of the wave are too quick for the magnet field inside a transformer so for a short moment the magnet field works against the current which is already coming from the other side. This creates heat inside the transformer and also disturbing electromagnetic fields. Digital appliances may be confused and sometimes totally unable to work because of this.

Trapezoid Wave Inverters

The output looks more like a pure sine wave and is suitable for use with more demanding appliances such as TVs, HiFi systems, inductive motors like those in washing machines, concrete mixers, etc. Trapezoid wave inverters are only slightly more expensive than the rectangular type.

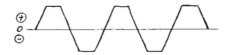

Most newer models of this type of inverter transform the electricity in two steps: First the 12V direct current polarity is switched back and forth (using power Mosfets) with a frequency of 30,000 to 50,000 Hertz by an electronic device. This alternating current is transformed with a very small transformer into a high tension. With a rectifier (4 diodes) it's changed again into a direct current of about 270V and stored in a big capacitor. Now in a second step, this high tensioned direct current is switched with another electronic device (and a second set of high voltage Mosfets), into a 230V alternating current.

Pure Sine Wave Inverters

These are the best inverters. They make almost the same sine wave as the National Grid and cause the least disturbance in music systems, TVs, CB radios, etc.

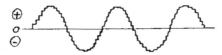

They work efficiently with appliances with built-in transformers and can be used with digital appliances such as computers, and sine wave sensitive devices such as videos recorders. They either have a costly electronic only on the DC side and a big and heavy transformer, or they work more like the trapezoid wave inverters with a two step method, but the electronic for the second step tries to rebuilt the sine wave in a digital way. There is a whole range of pure sine inverters of varying quality available, basically you get what you pay for, usually the more expensive the purer the sine wave.

In General

Many appliances take a 5 times higher current to start than they use in normal running! e.g. lamps, fridges, grinders, TVs, etc.

An 80W TV will need an inverter of about 400W to have enough power to start it. Some inverters have a Soft-Start-Device which is helpful.

The output Mosfets are very sensitive to disturbing signals coming back from the appliance into the inverter. When the brushes inside the motor of a drill are worn or dirty they make sparks. These sparks make a range of disturbing frequencies which inverters don't like at all. They can cause the output Mosfets to burn out, so make sure your appliances are in good condition.

There can also be a problem with low energy light bulbs. They are actually small fluorescent light bulbs, and they need to transform the 230V / 50 Hertz electricity into a much higher frequency. They run with a frequency of about 30,000 Hertz which can confuse some inverters.

Better quality inverters have a built in Net Filter to filter these disturbing frequencies out a bit. They are helpful but they can't filter everything away, just lower the disturbing frequencies.

Nearly all inverters have a built-in deep discharging regulator, which is very important because inverters must be connected direct to the battery. Inverters require so much current (1,000W = 120A!) that the deep discharging regulator (often only 10A) from a solar system is not able to cope. Make sure you connect the battery and the inverter with very short and thick cables!

The disadvantages of inverters are: the low efficiency rate (70% - 90%), the high initial cost, the danger to life because of the high tension (230V), the radiation of 'electrosmog' (see next chapter), which is why many people prefer to run most of their appliance direct on 12V rather than use an inverter (see also the chapter, Tips and Tricks, page 102). It's easy to have all your lights and music systems direct on 12V DC and so you only have to use the inverter with 230V AC occasionally for the TV, computer, food mixer, washing machine, etc.

Electrosmog

Mankind often has problems in the way it uses technology due to a mixture of ignorance and greed. It seems that it is only when confronted face to face with the effects of the associated dangers that you appreciate them. The transportation of data and the every day use of electricity which use electromagnetic waves is growing exponentially. Electrosmog caused by the large amounts of electromagnetic and electrostatic fields present, is coming into contact with people, animals and plants in increasing doses.

As in medicine and some other fields of science the actual effects of electrosmog are difficult to quantify due to the infinite amount of potential influences on the body, e.g. diet, exercise, genetics, etc.

Because research on the effects of electrosmog has proved contradictory, the difference of opinion has given rise to two groups. Those who believe it has no important effect and those who believe it is potentially extremely dangerous.

In modern science, such as medicine, you have to realize that so many things are connected that you have to take an objective overall view.

Electrosmog can influence us biologically, for example in our organs' regulatory systems, due to its constant presence, like a repetitive injury.

It is considered that insomnia, headaches, migraines, nervousness, paranoia, irritation, depression, stress, dizziness, and even acid rain and tree illnesses may at least in part be attributed to electrosmog.

Our bodies have many subconscious regulatory systems which have their own frequencies in different ranges. These systems can resonate with natural occurring frequencies as well as the technological ones and can store them up. As you already know, nature transmits and is controlled by these frequencies, in fact everything is in contact with IME (information carried by micro energy).

73

Electrosmog can be separated into electromagnetic and electrostatic fields which are both either alternating or constant (direct). As in current (AC, DC) these fields can be separated into low or high frequency ranges.

Electrostatic Constant Fields

These are electrostatic charges with no frequencies. You encounter these charges with synthetic carpets and pullovers which discharge making small sparks or when you comb your hair with a plastic comb making your hair stand on end. They are also present on television and old computer screens. In nature electrostatic fields discharge in clouds causing thunder and lightning.

With a simple voltmeter, e.g. a multimeter set to its DC range, you can measure these charges. There will be a reading in the display which, on touching metal objects with the probe, will go away after a few seconds.

Electromagnetic Constant Fields

When direct current is present in a conductor it emits electrostatic constant fields, e.g. in a battery storage system, digital watches and torches. The magnetic field of DC is similar to that of the earth but the most important difference is the direction of rotation of the electrons which is called Electron Spin. Magnets have these constant electromagnetic fields (EMFs). They are found in headphones, microphones, loudspeakers, telephones, electro motors... Any ferrous or nickelled metal can acquire these fields e.g. spectacle frames and welded bits of metal like on beds, bicycle frames and cars. You can locate them with a compass.

2° Variation	No risk
10° Variation	Weak field
100° Variation	Strong field
over 100° Variation	Extreme field

Electrostatic Alternating Fields

Mains electricity has a frequency of 50 Hertz which living organisms are sensitive to and emits ESAFs. In the low frequency range of up to 100 kHz (100,000 Hertz) ESAFs are caused by alternating current in conductors even when the circuit is not closed and the electricity isn't actually flowing. Transformers have strong radiation fields and also have a strong high frequency effect.

The ESAFs in high voltage pylons radiate a long way because of the high tension. The official recommended distance to stay away from a 380kV pylon is 127m and 22m from a 110kV pylon, but with a simply electrosmog meter you can hear them kilometres away! Trees and thick stone walls can help protect from these fields.

Solar power systems can also have these intruding ESAFs caused by electric motors, dimmer switches, most solar regulators (shunt regulators) and all inverters, so they can get to every cable in the system all around your home.

Measuring the ESAFs present in the body by linking yourself up to a multimeter. Recommended limits:

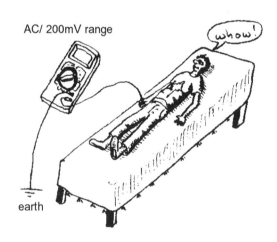

AC/ 200mV range

whow!

earth

In bed: 20mV
Living room: 500mV

To test the charge present in the air set the multimeter to 2V-AC. For ground connection you can use the copper central heating pipes or alternatively extend the earth wire to a metal pole of 1 meter in length, hammered into the ground. Make an aerial similar to that in the picture.

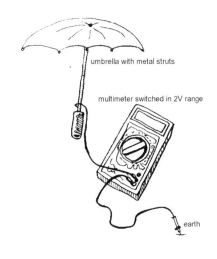

umbrella with metal struts

multimeter switched in 2V range

earth

There are certain systems available that send out a small DC voltage, waiting for something to be plugged in. This then quickly switches on the AC supply thus avoiding any unnecessary ESAFs.

Appliances such as televisions often use electricity when switched off or in standby mode. So it's always advisable to switch them off at the wall socket. If there is no switch here it is very simple to fit one to the *live* wire before the plug. But of course simply unplugging it works.

Electromagnetic Alternating Fields

These are caused by alternating current in conducting materials. Large currents make strong fields. This is particularly bad in inverters because the DC circuit involved attracts the alternating frequencies which causes very strong fields. This also happens with halogen lamps that use transformers because the field is very strong due to the conversion of tension causing the current to be much higher.

Telephone tapping machines can be used to pick up these frequencies but only in the audible range (20-20,000 Hertz). These machines can be bought cheaply, complete or in kit form in electric shops. Or you can make your own by using a guitar amplifier and a coil. Coils out of the small motor of the program switches of old washing machines work really well.

A coil with 1,000Ω and 2,500 windings is advisable, e.g. a guitar coil with the magnets removed.

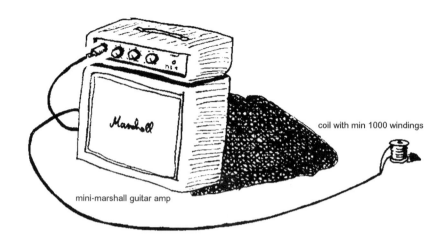

coil with min 1000 windings

mini-marshall guitar amp

With these you can hear EMAFs and also ESAFs by putting it in the vicinity of the field of e.g. energy-saving light bulbs, fluorescent tubes, transformers, inverters, TVs, computers, electric motors, dimmers, fridges, electrical blankets, electric stoves, household wiring, telephone systems, mobile phones, music systems, fuse boxes, sub stations and electric pylons.

With a multimeter on the 200mV-AC range you can measure EMAFs.

The High Frequency Range

These travel through the air and can get picked up on the low frequency waves of the national grid. Disturbing high frequency waves made by radio and television transmitters, mobile phones, radar stations, satellites and also quartz crystal clocks could disturb, for example, electronic injection systems in cars, or confuse washing machine programmes. Computers are very sensitive and aeroplanes have to go through very stringent tests in order to ensure they are well protected from this sort of interference.

The higher the frequency the smaller the amount of energy required for things to resonate. In human beings, most of the subtle psychological and biokinetic systems use these high frequencies (Mega and Giga Hertz, MHz and GHz). These frequencies are mostly immeasurable as the 'white noise' emitted by the technical measuring apparatuses is more intense then the frequencies emitted by the body. Because of the body's regulatory systems we are all highly sensitive receivers. Some people can sense these fields by holding divining rods and by dowsing. In principle everything is able to resonate so we always have an amazing number of factors coming together.

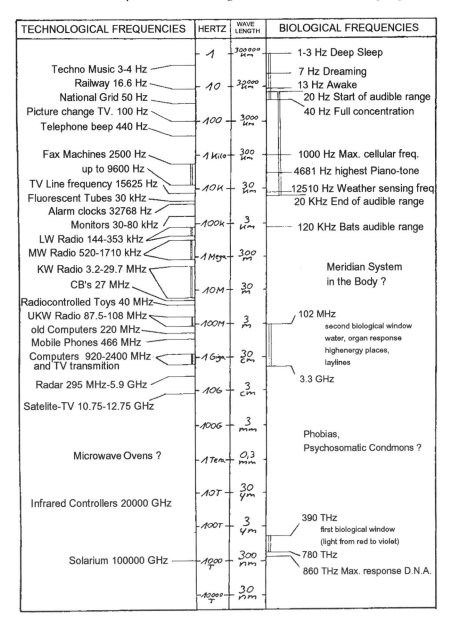

TECHNOLOGICAL FREQUENCIES	HERTZ	WAVE LENGTH	BIOLOGICAL FREQUENCIES
	1	300000 km	1-3 Hz Deep Sleep
Techno Music 3-4 Hz			7 Hz Dreaming
Railway 16.6 Hz	10	30000 km	13 Hz Awake
National Grid 50 Hz			20 Hz Start of audible range
Picture change TV. 100 Hz	100	3000 km	40 Hz Full concentration
Telephone beep 440 Hz			
Fax Machines 2500 Hz	1 kilo	300 km	1000 Hz Max. cellular freq.
up to 9600 Hz			4681 Hz highest Piano-tone
TV Line frequency 15625 Hz	10k	30 km	12510 Hz Weather sensing freq
Fluorescent Tubes 30 kHz			20 KHz End of audible range
Alarm clocks 32768 Hz			
Monitors 30-80 kHz	100k	3 km	120 KHz Bats audible range
LW Radio 144-353 kHz			
MW Radio 520-1710 kHz	1 Mega	300 m	
KW Radio 3.2-29.7 MHz			Meridian System
CB's 27 MHz	10M	30 m	in the Body ?
Radiocontrolled Toys 40 MHz			
UKW Radio 87.5-108 MHz			102 MHz
old Computers 220 MHz	100M	3 m	second biological window
Mobile Phones 466 MHz			water, organ response
Computers 920-2400 MHz and TV transmition	1 Giga	30 cm	highenergy places, laylines
Radar 295 MHz-5.9 GHz	10G	3 cm	3.3 GHz
Satelite-TV 10.75-12.75 GHz			
	100G	3 mm	Phobias,
Microwave Ovens ?	1 Tera	0,3 mm	Psychosomatic Condmons ?
	10T	30 µm	
Infrared Controllers 20000 GHz			
	100T	3 µm	390 THz
			first biological window (light from red to violet)
Solarium 100000 GHz	1000 T	300 nm	780 THz
			860 THz Max. response D.N.A.
	10000 T	30 nm	

Electrosmog In 12V Power Systems

Some regulators can give off a disturbing field between the frequencies of 1Hz to 1,000Hz, especially those with 'pulsed current regulation' when the battery being charged is full. Take care also with electro motors, dimmers, inverters, quartz crystal clocks, TVs, computers and mobile phones. Solar systems tend to be safer as the appliances are not constantly switched on. A lot of the main appliances in solar systems do not make alternating fields, such as lamps, batteries and solar panels.

It's all much better than the mains which attracts many other frequencies. The biggest danger is to be exposed to constant frequencies over prolonged periods of time, such as the 50Hz in mains electricity. But by far the worst is high frequencies, such as in mobile phones.

The table on page 78 shows the spectrum of frequencies with some technical and biological frequencies. Look at it bearing in mind waves don't just resonate at their fundamental frequency. they produce many 'overtone' frequencies. The first of these being twice the speed of the fundamental (one octave higher in music) then 3 times, then 4 etc.

For example a wave of 50Hz like in the National Grid system will produce overtones with 100Hz, 200Hz, 400Hz, 800Hz, 1.6kHz etc. And they even can make other systems using one of these frequencies resonate to it.

Lamps

In most solar systems, lamps are the most common power users. A good, efficient and economic lighting system is very important, partly for the longevity of the battery but also for the fun of running a solar system.

Using small lamps of 5W or 10W, in specific areas, e.g. table lamps, bedside lights, etc., is more efficient than one big light for the entire space. Pale coloured walls and mirrors reflect light and help provide general illumination. The following lamps work directly with 12V:

Car Lights

These can be 2W, 3W, 5W, 8W, 10W, 21W, 45W (or 55W if they are halogen bulbs). You can use second hand indicator sockets, or interior light sockets, which can be bought cheaply from your local scrap yard! You can also solder the cables directly onto the lamps with a strong soldering iron. As a guide a 3W lamp is a little brighter than a candle.

Reflectors direct the light to the area needed. Sometimes you need only 30% of the power of the lamp if it's well directed. A 5W lamp with a reflector can give as much light as an undirected 15W lamp. Reflectors from car headlights, torches and bicycle lamps are sometimes too exact, focusing the light into too small a point. You can dent metal reflectors by hitting them all over with a small ballpane hammer, thus increasing the surface area and diffusing the light. Homemade reflectors can be made from the concave bottoms of pressurized tins, e.g. beer tins, aerosol paint cans, small gas cylinders, etc. Make sure they are really empty first!

Halogen Lamps

These have a higher element temperature, so they are brighter, and are more efficient. They use the 'noble gas' halogen in the bulb to keep oxygen away from the element, so they last longer, up to 3,000 hours (normal lamps have a working life of 1,000 hours). So halogen lamps are the best choice for use in a 12V solar system.

You can buy them in any hardware shop: as bulbs of: 5W, 10W, 20W, 35W, 50W, 75W and 100W, or reflector lamps of: 10W, 20W, 35W and 50W with reflector diameters of: 35mm or 50mm, in different types: spotlight (12°) or floodlight (30°). Floodlights are usually the best choice.

A halogen lamp gives out a very large spectrum of light including the ultra violet (UV) range. You can buy bulbs made from UV safe glass, or with a UV safe coating, but these are only really necessary for lights which will be shining onto food, for example in shop displays, as the UV rays can change the colour of the food. The UV safe glass filters out the UV rays, but there is no danger to humans from halogen lamps; the UV light produced is a small percentage of what we get naturally from the sun.

You can also buy reflector lamps with special infra red (IR) safe reflectors, which filter the IR range of the light out through the back of the reflector: red light can be seen at the back of the lamp, but it isn't dangerous. They are used because IR rays heat as well as light, which is not practical for lighting objects which must not become warm, for example fresh food. These lamps, sockets and surrounds can become hot where the IR rays are being filtered out of the back. Because the red spectum has been cut the light produced appears cold and rather harsh. You can recognize a lamp with an IR safe reflector from the packaging and from the colour of the reflector which appears blue or green, as opposed to silver.

Halogen bulbs get very hot, especially bulbs of 20W or more. You should use proper halogen lamp fittings because of the heat. Don't use plastic choc blocs to connect to the bulbs as they can melt and create a short circuit by the positive and negative touching together. Grease from fingerprints on the bulb can carbonize and turn black, allowing less light through so handle with care.

Because of the higher temperature of the element halogen lamps are more sensitive to vibrations than normal car lamps. If you knock them or drop them when they are on they often break.

When a reflector lamp bulb is broken you can replace it with a single halogen bulb, which is cheaper than buying a whole new unit. For this you must remove the safety glass, if any. Sometimes you have to break it, and then gently tap the bulb out from the back with a small hammer. The bulbs are usually set into the reflector with plaster.

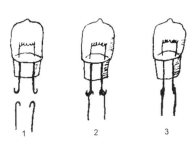

You can then fiddle in a new bulb which will hold well enough even without the plaster. If you want to put a new 5W or 10W bulb in a used 50mm reflector you will need to make the wires longer. The wire from the bulbs are made of stainless steel so they can't be soldered with normal soldering stuff.

You can get over this by taking two solid copper wires and looping the ends. If you use 5W or 10W bulbs you can try to carefully loop the ends of the two stainless steel wires. Then hook the loops together and pinch them tight. Afterwards you can solder the copper for a better connection.

If you use bulbs of 20W or more it's better not try to bend the stainless steel wires, because the glass cracks very easily. Wind the solid copper wire around the stainless steel

wire or bind both wires together with a very thin copper cable and then solder it.

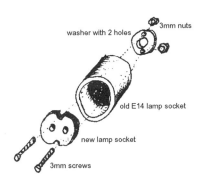

washer with 2 holes
3mm nuts
old E14 lamp socket
new lamp socket
3mm screws

A halogen lamp socket fits into a normal 230V lamp socket (type E14). You can fix it inside with two 3mm screws and a washer with two extra holes drilled in it. So its quite easy to change this type of 230V lights into 12V.

Energy Saving Lamps & Fluorescent Lamps

These are more efficient than halogen lamps, but make some electrosmog because both of these lamp types have a transformer and run with alternating currents of around 30,000Hz. In energy saving lamps the transformer is built into the lamp socket, and in fluorescent tube lights the transformer is mounted in the box fitting. The light produced can look cold and uncomfortable because the red range of the light spectrum is very poor.

Bicycle Lamps

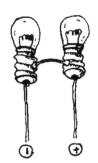

As we mentioned in the chapter Watts and Volts you could make a 12V lamp from two 6V lamps with the same power.

Power LED Lamps

New, super effective! Usually only used in torches and solar powered garden lights, but now also available in 12V, like the halogen reflector lights, but with a lot of LED (Light-Emitting-Diodes) inside. They use very little energy (like 1.5W) and in comparison to normal light bulbs they give out a lot of light. But the light colour is usually very cold and uncomfortable. However there are new LED types available which emit a nice and warm light, instead of the hard bluish ones..

See the chapter, Tips and Tricks, page 112, for how to make torches with these LEDs.

Music Systems

These are the second most common use of solar power. To avoid using inverters you need to find a system that runs on 12V. You can use car radio cassettes / CD players and car power amps, but car systems use a lot of power; for example a car radio can use 10W of power just by being switched on.

More economic systems are portable radio cassettes or 'ghetto blasters', because they are built for using battery power. Some of them use eight batteries, each of 1.5V, totalling 12V. These radio cassettes can be wired to our 12V solar system with a simple connection inside the battery box:

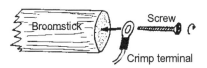

Using a broom handle cut to the right length, attach a cable to the flat face of the handle, using a screw and a crimp terminal.

The springs in the battery box always connect with the minus terminal of the 1.5V batteries. The flat metal plate in the box connects with the plus. It's very simple when the batteries are arranged in a single line. If not you have to check the method of connection: There will be just a simple connection between plus and minus at one end, and the

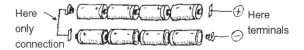

terminals (spring and plate) at the other. If it is not obvious what's going on, measure using the Ohm (resistance) range of the multimeter (see the chapter, Multimeters, page 47).

Instead of the batteries, you put the broom handle in the battery box, making contact between the terminal in the radio cassette and the crimp-terminal/screw in the broom handle.

Make sure that you don't confuse the poles, by using a diode and/or a special plug (see the chapter, Plugs & Polarity, page 56).

It's very useful to put a diode in, the loss of tension is usually no problem.

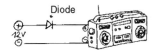

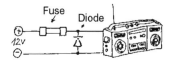

But if the radio runs with 10 batteries of 1.5V = 15V it's better to put a diode in which blocks when the polarity is connected correctly, but when the poles are confused it opens and lets a fuse blow.

But most smaller 'ghetto blasters' use six batteries of 1.5V, totalling only 9V. However, when you use 9V appliances with the 230V of the National Grid, the voltage of the transformer inside is usually more than 9V, so it follows that you may be able to use these 9V appliances on a 12V system without damage. But when the sun shines on a solar system the battery voltage can increase to 14V! That's quite a lot for a 9V appliance. Most of the time it doesn't matter, but if you want to

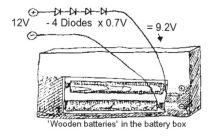

'Wooden batteries' in the battery box

be sure that running your precious 9V radio cassette on a 12V system is totally safe, you can put in a series of diodes. With the loss of tension of 0.7V per diode, the voltage will decrease 2.8V if you use 4 diodes, e.g. from 12V down to 9.2V. The disadvantage of this is that the diodes will decrease the voltage by 2.8V, regardless of the battery voltage.

$$14.0V - 2.8V = 11.2V$$
$$10.8V - 2.8V = 8V \text{ only}$$

Use four diodes for 9V appliances and 7 or 8 diodes for 6V appliances.

A better method is to use a voltage regulator, which regulates the tension exactly, at every battery level. These need to be cooled with some form of heat sink. You can buy them cheaply in electronic shops for the following tensions: 2V, 5V, 6V, 7.5V, 8V, 9V, 10V, 12V, 15V, 18V and 24V. They have a typical serial number: 78 and then the tension so they are usually named: 7809 for 9V (1A) and 7812 for 12V (1A), or 78S09 for 9V (1.5A) and 78S12 for 12V (1.5A).

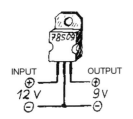

INPUT ⊕ 12 V ⊖   OUTPUT ⊕ 9 V ⊖

They are connected like this:

Here is a more luxury version of the circuit. Don't forget to screw a heat sink on:

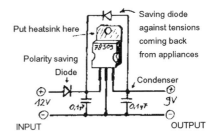

Saving diode against tensions coming back from appliances

Put heatsink here

Polarity saving Diode

Condenser

INPUT ⊕ 12V ⊖   0,1ᵧ𝄼   0,1ᵧ𝄼   OUTPUT ⊕ 9V ⊖

If there is enough room in the battery compartment, you can put a diode in to make it absolutely polarity-safe and put the voltage regulator with a heat sink onto the broom handle, too.

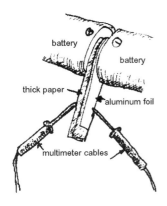

battery

battery

thick paper

aluminum foil

multimeter cables

The S-Version with max. 1.5A (e.g. 9V x 1.5A = 13.5W) is strong enough for most 'ghetto blasters'. The maximum power marked on the product is usually overstated. A maximum power of 5W-24W is normal, even if it's marked 200W. Diodes and voltage regulators with 1.5A maximum power work well in most cases if they have a big enough heat sink, but if you are not sure it is better to measure the maximum current (at full volume) with a multimeter in the 10A range.

You can do this with a piece of card, coated on either side with aluminium foil. Place this between two of the 1.5V batteries in the box, so the circuit is broken. By connecting the cables of your multimeter, one to either side of the card, you create the circuit again, running through the multimeter and measuring the current at the same time. It's interesting to test the amperes used at low and high volumes, and also in standby.

Some manufacturers products, e.g. Sony, need two different tensions (e.g. 9V and 4.5V) otherwise they just don't work or they show something in the display like 'battery fault'. They have an extra output at a connection in the battery compartment, usually in the middle of

86

the battery line. The only way to adapt these to your 12V solar system is to create these tensions with two voltage regulators and connect them correctly at the right points in the battery compartment.

It would be very smart to make a separate 12V connection with a diode and a voltage regulator inside the body of the product. But to do this you must open the case up and you will invalidate all guarantees on the equipment by doing this:

Put the connection from the solar system onto the big electrolytic capacitor behind the transformer and the four diodes (rectifier).

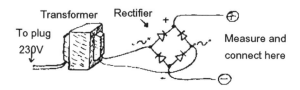

While you have the radio cassette open you can also measure the output voltage of the transformer.

And check the polarity before you make the connection. There is always a sign on every electrolytic capacitor on the minus pole.

You can do this kind of conversion on many 230V appliances which have a transformer inside. If the voltage is around 12V, you can use it directly with your 12V system.

And if you want to use your music system only on 12V, you can take away the transformer and make the connection where the transformer was connected before. The four diodes then work like polarity guards, so it doesn't matter if you confuse the poles because it now works in both directions.

Cordless Tools

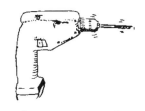

There are more and more cordless tools, which in principle are usable with the direct current of a 12V system. The most common is the cordless drill, but you can also get cordless jigsaws (Black & Decker, Atlas Copco), sanders (Metabo) circular saws (Parkside, Mannesmann, De Walt), hot glue guns (Metabo), delta sanders (Skill), angle grinders (Metabo), engravers, etc.

The company Proxon has brought out a small range of 12V tools (drill, milling machine, jigsaw, angle grinder, sanders, soldering iron...).

In auto shops you can buy 12V vacuum cleaners, coffee machines, ventilators, compressors, small water boilers, irons and hair curlers.

Usually you need more time and patience to work with 12V tools, because they have much less power than 230V tools. But they are more economical, and so better to use in small solar systems.

The DC-electro motors in the tools work with a very wide range of tensions, so it is not a problem to use them with 12V even when the voltage of the power pack is higher or lower. Tools of one make often come with different power packs (e.g. 7.2V, 9.6V, 12V). The tools all have the same motors and speed-electronics, just the power pack, recharging unit and labelling are different. It is possible to use cordless tools with voltages which range from 4.8V to 18V directly on 12V.

But beware, the motors of low voltage tools will get hot more quickly when they run with higher tensions and can burn out if they overheat. Let them cool down after prolonged or heavy use . Higher voltage tools will just work with less speed and less power.

Even the electronic speed governors work in a wide range of tensions, and can be used with 12V.

Some companies also sell tools without power packs (Black & Decker), which is ideal because you don't need the power pack and recharge unit. You can buy tools with broken powerpacks very cheaply in flea markets.

Mark the polarity!

Short Screws

Crimp terminal

Wood Block

To connect up your soon-to-be corded 12V tool, you can use the plastic case of the power pack. Take the rechargeable batteries out and solder cables to the metal plates. Alternatively make a wooden block of the same size as the power pack with power terminals in the same position. Use very short screws (otherwise there's a danger of short circuit!) and crimp terminals to secure the cables to the block.

When you have tools with a power pack fixed inside, you must open the machine and take the pack out, otherwise they would overcharge when you connected them to the 12V system.

Cordless tools which have motors without speed governors are not polarity sensitive. They just run backwards if you connect them the wrong way round. However this is very dangerous with circular saws, etc.

Cordless tools with speed governors on the other hand will be destroyed if you confuse the poles. You can tell if the tool has a speed governor: the power builds up as you increase the pressure on the button or trigger; you can also hear a high pitched whining noise before the machine reaches full power.

Mark the polarity clearly on the wooden block and tool, so as not to insert it the wrong way round. If you don't know the polarity look for marks of (+) and (−) on the power pack. If you don't find anything you can always measure it using a multimeter. In the absence of a power pack , open the machine and look for clues, e.g. cable colours: red(+) and black(−), or marks on the internal connections e.g. on the speed governor B+ and B− for the battery and M+ and M− for the motor. The motor itself is sometimes marked with (+) or (−), too.

To avoid confusing the poles you can solder the cables directly onto the contacts inside the machine, and use polarity-safe plugs.

If you use long and thin cables, e.g. 2x 0,75mm², to run a 9.6V cordless drill, the cables will get a bit warm. But the voltage will drop down when you use a lot of power (drills up to 25A!), the drill won't be overpowered so easily and will not get hot so quickly.

Solar Grinders

You can make a 12V solar grinder by using an old windscreen wiper motor from a scrap yard. These motors are usually around 100W (about 8A) and still in good condition. Grinders made with them are less powerful than normal mains grinders; so you need more time and patience when using them. On the plus side tools being sharpened 'glow out' (and lose their tempered strength) less quickly, and the grinders also use less electricity.

You will need the following material:

- An old windscreen wiper motor.
- A grinding stone of 100 or 150mm diameter.
- Small block of hard wood, to cut down to place between the axle and the centre of the stone.
- Wood to make a case, guard and a hand/tool rest.
- Two big washers, middle diameter: 6mm, exterior diameter: greater than that of the hole in the grinding stone. You could cut these yourself from metal sheet.
- Two 6mm nuts and a handful of countersunk screws.
- A switch, some cables and a crimp terminal.
- Insulating tape or choc bloc connections.
- 10cm of thick rubber tube, (e.g. car radiator hose).

And the following tools:

- A woodsaw and a hacksaw.
- A good file, a screwdriver and a micrometer.
- A chisel, a hammer and a vice.
- A drill with a 6mm bit.
- A 6mm thread cutter and a pair of pliers.
- A 10mm spanner.

Windscreen wiper motors have a 'worm drive' gear system which you don't need, so you have to dismantle everything, then cut off the case of the gear system (found at the front) with a hacksaw. Clean and re-grease the two bearings, and then put it all back together again.

Hold the motor in the vice and connect the battery minus cable to a screw on the body of the motor, using a crimp terminal. Windscreen wiper motors have different speeds: usually a red cable is for fast and a green for slow, so connect the battery plus cable to the (red) cable for the fast speed. Insulate off the cable you are not using.

You have to file off the thread on the axle with a good file, so let the motor turn and file against the direction of rotation, until you reach a smooth diameter of exactly 6mm. This will take a bit of patience and dedication.

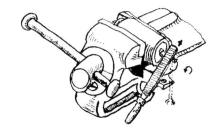

Next you need to cut your hardwood block. This fits exactly over the axle and into the hole in the centre of the grindstone. It acts like a spacer, but more importantly it ensures a smooth and central rotation for the stone. The hole you drill in the centre of the wood must fit snugly over the axle, and it's easier to make an exact round hole if you follow the grain of the wood. Bearing this in mind, cut down the outside of the wooden block until it is a little larger than the internal diameter of the grindstone. Drill the 6mm hole in the centre, then fit the wooden

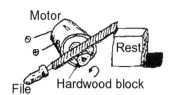

block over the axle. Turn on the motor and take a file again and file down the wood until it is exactly the same size as the hole in the grindstone. If you steady the end of the file on something solid, you achieve a better round finish.

Remove the block of wood from the axle again, and holding the axle with the pliers, cut a new thread at the end. This should be long enough to screw the 6mm nut onto; or if the axle is fairly long, you may have enough space for a nut with a locking nut as well, so make your thread longer accordingly.

Assemble the washers, woodblock, grindstone and nut(s) like this:

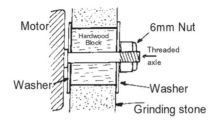

If you have just one nut you can paint the thread just before you put it on. When the paint dries, it will 'glue' the nut to the thread for extra security. Turn the motor on, and check that the whole assembly turns smoothly and centrally, without too much vibration. If all is well you can continue, but if not, it is likely that the wood block wasn't well enough cut, and you must make another. Don't give up! It will be worth it...

A wooden case will protect the grindstone, and also yourself from flying fragments of metal and stone, which are very dangerous for your eyes! Always wear safety goggles when working with grinders!

Make four 'feet' for the case to grip the ground, and prevent it from 'walking' with the vibration produced.

To make a rest to steady your hand or tool whilst sharpening, saw and chisel out a slot in another block of wood, just wide enough to place the grinding stone into. Attach the switch to the apparatus, and it's ready.

Happy grinding!

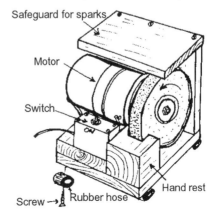

Solar Circular Saws

Circular saws need more powerful motors (e.g. 1,000W and more). 12V electric motors in this class include car and motorbike starter motors. The motorbike ones are best as they are more economical for the power output, especially those with permanent magnets inside. You can usually feel these when you turn the cog.

A solar powered circular saw is relatively simple to make using an old drill-powered circular saw, which you still find in flea markets and at garage sales, etc.

Motorbike starter motors are very small and handy. Most of them have the same round shaft with a small cog on the end.

To make the join between the motor and the saw you need two sleeves of metal. You will probably need to ask some-body with a mechanical work-shop to do this for you...

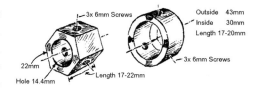

- 3x 6mm Screws
- Outside 43mm
- Inside 30mm
- Length 17-20mm
- 3x 6mm Screws
- 22mm
- Length 17-22mm
- Hole 14.4mm

The first, larger one attaches to the shaft of the motor and the outside of the saw socket. The smaller sleeve fits over the cog and goes inside the saw socket. Use three small grub screws to fix the sleeves to the shaft and the cog.

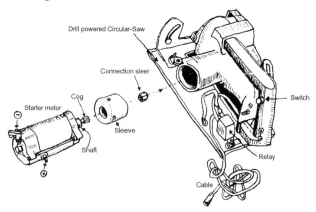

- Drill powered Circular-Saw
- Connection sleer
- Cog
- Starter motor
- Switch
- Sleeve
- Shaft
- Relay
- Cable

93

Use a big relay and a small switch, e.g. doorbell type, on the handle of the circular saw, to switch on the power from the battery to the motor.

To prevent the motor overheating, you can use slightly thinner cables than normally allowed, these decrease the maximum revolution rate of the motor, e.g. for currents of 100A use 4mm^2 cables of 1.5m length. The cables will get quite warm, so you must use good quality and heat safe insulated cables!

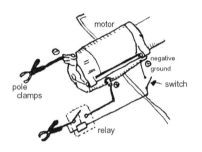

Use pair of big crocodile clips from good quality jump leads to connect to the battery.

But you will still only be able to saw for a short time (max. 2 minutes!) until the motor gets too hot. These motors are made to be used for a few seconds, to start the engine of the motorbike, so their construction is such that they can't get rid of heat quickly. You can improve this by drilling some small holes into the front and the back and fitting a cooling fan onto it. You can fix the whole thing under a table to get a 12V table circular saw... And... Sharp blades saw much better!

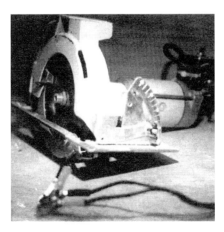

Starter motor powered portable circular saw.

The same saw mounted under a cutting table.

Sun-Following Systems

With a simple sun-following system you can get almost 1.5 times more energy from your solar panel. To make a really simple one, you need a pallet for the base, a bicycle front wheel, a piece of thick metal plate (to attach the wheel onto the base), a small solar electric motor, like the ones you

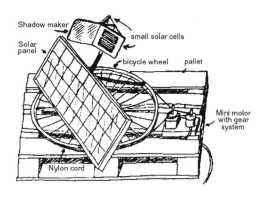

find in solar experimental kits for kids, or any other low voltage motors like a walkman motor, motors out of a broken CD player, or old cameras, and you need four (or sometimes six, depending on the motors) and four small solar cells. The motor/s must be able to run from one pair of cells in series.

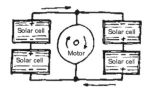

You connect two pairs of solar cells like this:

Each pair is connected the opposite way to the other. If both pairs get the same light, the electricity runs in a circuit and the motor doesn't turn.

If there is a shadow on one pair of the cells they will block the electricity and the electricity from the pair in light now runs the motor. One pair of cells turns the motor clockwise and the other turns it anticlockwise.

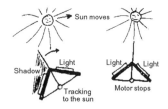

The 90° angle between the cells helps to turn it back in the morning.

The motor connects to and turns the wheel via a nylon cord or a band of rubber. To set up, attach your big solar panel and the arrangement of the two pairs of small cells onto the bicycle wheel. Place a small thin sheet of metal upright between the two pairs of cells to create a shadow with the movement of the sun. The system runs without using any other electronics.

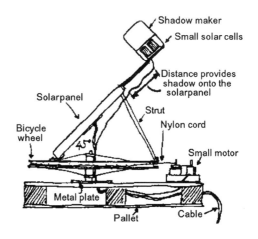

If the motor has a gear system (as with some camera motors) you can also attach it the opposite way (see pictures below), the wheel is fixed to the pallet, and the motor is turning together with the solar panels on the T-frame which is fixed on the axle of the wheel. Brake pads are placed very near to the wheel to stop the whole thing when the wind tries to turn it.

Nifty, eh?

Above: Geared motor on frame turns against a fixed wheel.

Left: Bicycle wheel solar tracking system mounted on a pallet.

Solar Welding

There is no magic in welding with batteries, it's actually very easy. You only need three car batteries (all sizes from 38Ah to 120Ah). Don't use special solar batteries, they can't give out enough current and would probably be damaged if you tried. You have to connect them in series connection to get a tension of 36V. Even when you use small like 44Ah you will still be able to weld for some hours, but only when the batteries still have enough capacity. If you use old batteries it's better to double things up by connecting pairs in parallel (you will need 6 batteries all together).

It's very important with these high currents of up to 200A that you use very thick cables (min 15mm² though 50mm² is much better): like good jump lead cables, battery connection cables from a car shop, or welding cables from 230V welding equipment.

Equally important is good contacts on the battery poles. Use only proper high quality pole clamps, not crocodile clamps, and clean the poles and the pole clamps before use with hot water and a metal brush or sand paper.

Connect the welding electrodes to the minus (–) pole of the first battery, and the metal you want to weld to the plus (+) pole of the third battery.

Car batteries can easily give out up to 300A. But that's more than you really need to weld, the electrodes would get too hot and the welding area too big. So you have to bring the current down with a simple resistor of fence wire wound around something heat resistant and non conducting like a brick.

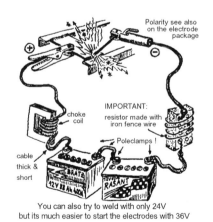

Iron fence wire regulates the current perfectly. If it's cold it lets all the current flow, but if it gets hot the resistance increases very quickly and it regulates the current down.

If you want to make long welding seams, use fans to cool the fence wire otherwise the current flow will become unstable.

For electrodes with 2 to 2.5mmØ use 6m of 3mmØ fence wire. The exact length varies with the age and capacity of the batteries and the current you need for welding (50-150A). Make some connection points on the fence wire at say 10cm/20cm/40cm/80cm/1.5m/3m/6m, then you will always be able to set the right current.

To charge up a 36V system, you would need three solar panels. To be able to charge the whole system with only one solar panel, the batteries need to be in parallel connection. Switch them back to series connection only when you want to weld, e.g. with four very big home made change over switches. See general connection plan below:

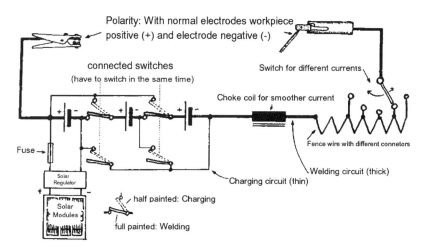

Very professional but not really necessary is a choke coil wound with big cable round the core of an old transformer. When a high current flows the magnetic field increases inside the transformer which gives back a push of current in the system for a few milliseconds and the current drops down. The welding seams produced will be cleaner and flow better.

A powerful workbench power supply which has three 90Ah batteries inside. Normally connected on 12V for using with tools like drills, saws, grinders, etc. out of the sockets. It has a built in regulator and can be charged with only one solar panel. Using the specially constructed changeover switch you can switch to 36V for welding. The built in volt meter shows the tensions of all three batteries or individuals via a multi pole change over switch. In the sewing machine box on the top is the current regulator, this can also be used separately with another set of batteries.

Here you can see the inside of the sewing machine box, a choke coil, two ventilators for the fence wire with a 5 pol. change over switch to set different currents (max.120A for big electrodes or thick metal to weld – and min. 50A for small electrodes and thin metal). In the base there is a place to store the welding cables.

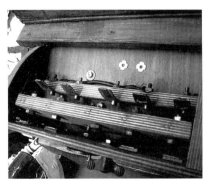

This is the 'magical' 4 pole change over switch for high currents, made from 10mm screws and metal plates. This allows the connection change over of the 3 batteries from parallel to series.

Battery Tester

If you want to sort out the best battery from those available at the scrap yard, or check out any battery properly, you need to make a special tester.

With a multimeter and an acid tester you can already find out quite a lot about a battery, but only when the battery is giving out a lot of current can you gauge it's true condition.

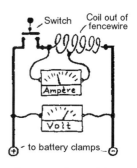

The battery tester shown here has a coil made from 2.5m iron fence wire with 2mm diameter. The big button switch connects this coil to the battery, and sucks out around 80A to 100A of current. To measure the battery voltage and the actual current press the button for a few seconds. You can measure a couple of times in quick succession but the coil will get quite hot, so let it cool down between tests. A homemade voltmeter (see next chapter) and an amp meter made out of an old VU meter from a tape recorder display the results.

Connect the plus(+) of the voltmeter at the end of the coil and place the minus(−) using the metal part from a connector block (the plastic would melt) to the place on the coil where you get a full reading when the current flows from a healthy battery. It would be quite complicated to calibrate the meter, but it's not really necessary to know the exactly amount of current, it's only important to see if the current stays at more or less the same level. Totally broken batteries collapse totally in both tension and current. When the voltage drops down below 9V the battery is very, very old, even if it can hold the current constantly. A good battery drops down only 1V to 2V.

Portable battery tester

Only use big pole clamps, thick cables and a special battery switch.

Special Voltmeter

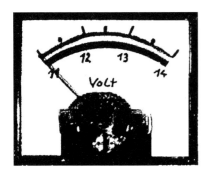

Voltmeters with a range between 11V and 14V are very practical for solar systems because in this range you can learn all about the normal capacity of your batteries. Analog voltmeters with this range are quite rare, but you can make yourself one out of an old analog multimeter or with a tape recorder VU meter.

The first step is to take it completely apart, being very careful with the very sensitive needle. Clean off all the scale markings with acetone, or glue some white paper over them. Then put it all back together again, except the lid. Connect a 10V Zener-Diode (from an electronics shop) to the plus(+) terminal with the ring on the diode pointing away from the meter. Then solder any normal diode (like IN-4148) on to it with the ring pointing towards the meter. Temporarily connect a 27kΩ to 100kΩ poty to the minus(–) terminal of the meter.

You now need a bench power supply (see the chapter on page 118). Set it to 14V and connect the meter to it. Next turn the wheel of the poty until the needle shows maximum. Then disconnect it from the bench power supply and measure the resistance of the poty with a multimeter in the Ohm(Ω) range. Now look for a resistor with exactly the same resistance and solder it onto the minus(–) of the meter.

Then calibrate the meter exactly by setting different tensions with the bench power supply (11V/11.5V/12V/12.5V/13V/ 13.5V/14V) and marking the position of the needle each time with a permanent marker. The consumption of these meters is so little (0.5mA) that you can leave them permanently connected.

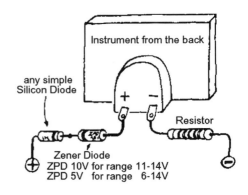

Instrument from the back

any simple Silicon Diode

+ –

Resistor

Zener Diode
ZPD 10V for range 11-14V
ZPD 5V for range 6-14V

Tips & Tricks

Repairing Old Batteries

Tip 1: Corrosion on the battery terminals can produce resistance. Clean them with a wire brush or sandpaper, and protect them with a smear of Vaseline.

Tip 2: A layer of dirt, especially damp dust on the outside of the battery can create a connection between the two terminals speeding the battery's self-discharge rate. If dust or dirt falls inside the cells, it can destroy the plates chemically.

Tip 3: Unequal cells? If you can see that the liquid levels are different between cells, you can charge each cell separately to equalize their capacity. Attach clean metal strips to a positive and a negative plate (test first with multimeter). But carefully! Don't make a short circuit by connecting positive and negative plates with your strip.

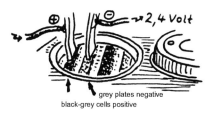

grey plates negative
black-grey cells positive

Put 2.3 or 2.4V through the strips into the battery cell.

A cell with a high liquid level (deep discharged) needs a small current for a long time until it reaches 2.4V.

A cell with a low liquid level (low capacity, overcharged) needs topping up with distilled water and charging normally. In the worst case, you can start by giving the cell a 'shock charge' of 6-12V (as in Tip 5, but only for one cell). Then discharge the whole battery and charge it up normally.

Tip 4: Sulphur build up. Symptoms: a light grey coating on the plates, a lower capacity and a lowering of the maximum current.

- You can use battery pulsers like the 'Megapulse' which cracks the sulphur layer up by very short pulses of a short circuit (only some milliseconds every 5 seconds with a current of up to 80A).

- Or use the battery very hard as in a welding system.

- Or you can charge the battery for a long time at 14.4V. In the worst cases, empty the acid out of the battery and refill it with distilled water. Charge it constantly for several days and nights! Then empty the distilled water out of the battery, refill with acid, check that the level of acidity is correct (using an acid tester).

- Simpler but rougher: open the tops of the cells and charge at 24V over a resistor (iron wire, see diagram, page 104) for a few minutes at a high current (up to 50A for a 100Ah battery). Then discharge fully with as high current as possible. Immediately after, charge up normally (from a solar panel with a regulator). See safety notice on next page 104!

Tip 5: Extreme sulphur build up: Symptoms: when you put 14.4 V through the battery, it hardly takes any current (A); the battery cannot give enough current for heavy loads; also the battery voltage collapses to under 9V when you try to take more current out than 50A (test with the special battery tester, see page 100).

A violent charging trick helps: a very high voltage breaks up the layer of sulphur on the cells. This is possible because some acid is still present between the lead plate and the layer of sulphur. The high voltage causes the acid to form bubbles which break the sulphur layer apart.

Risk: Small particles of lead could fall from the plates and create a high self-discharge (or even a short circuit) between the plates. But remember! Your battery wasn't working properly anyway, and this works in three out of four cases.

Method: Top up liquid levels before and after. Make a chain of fully charged batteries in series to produce 24V, 36V or more. Charge the old, broken battery with this voltage. If needs be, step the voltage up to

96V (= eight batteries in series), until a small charging current starts to flow (6-10A for a 50-100Ah battery). When the current has reached 6-10A disconnect. You can use a trip switch or circuit breaker, with an automatic fuse, to cut off the charge automatically at 6-10A. Lower the voltage by one battery (12V) at a time and reconnect. Continue until your battery takes current at 24V. Then refill and charge normally.

IMPORTANT! RISK OF EXPLOSION!!

- Place batteries outside, and connect the entire system to one switch. This switch must be inside, away from the batteries otherwise a spark from it could ignite the gas produced by the batteries

- Smoking and naked flames forbidden!

- To prevent cables melting, place fuses between each battery (16A for Tip5, 60A for Tip 4).

- Cover the fuse-holders securely with plastic, as sparks released by blowing fuses could ignite the gas produced by the batteries

- Ensure all batteries have good connections!

- Use thick cables, minimum 10mm^2.

- Open all batteries whilst charging, to release explosive gases.

- This system must be monitored and not left alone whilst connected.

- Do not even think about doing this anywhere other than outside!

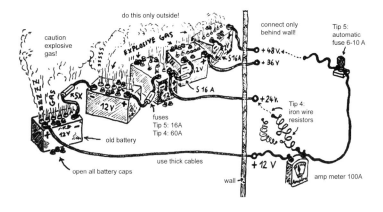

Gel Batteries

The gel can dry out, so no current will flow, as in batteries with sulphur build up. Open them up (ignoring the warning not to open!) and refill with distilled water, the same amount in every cell. Then charge normally. If the battery is still not good, the cells are probably suffering from sulphur build. Use Tip 4. but only with the maximum charging current allowed for this size and type of battery.

Tip: Many gel batteries have normal opening tops, with something stuck over them to prevent you opening them. You can prise this off with a screwdriver. When you open them and put water in, remember they can now leak like normal lead acid batteries. When the gel is very dry and the distilled water doesn't seem to penetrate, try using battery acid instead.

Important: Constant charging should only be 13.6 - 13.8V, not 14.4V because this will dry the gel out too fast.

Many gel batteries are fixed into place (e.g. in a wheelchair). Be careful not to screw them in too tightly, as the pressure can buckle the plates.

Also don't store them on top of each other etc.

Any mechanical pressure can cause gaps between the gel and the plates, where no current can flow. Tenths of millimetres can make a difference!

This can also happen when the maximum charging voltage is set too high. If the battery looks like it's pumped up, open the cell tops and press the battery carefully back into shape. Then top up by adding a uniform small amount of battery acid to each cell.

Repairing Solar Panels

Confused The Poles?

There are some people who still run their solar system without a regulator or a saving diode to stop reverse currents in the night. If you connect the solar panel to the battery the wrong way round the diodes inside the solar panel connector box blow up. Most of the time these diodes then create a short circuit. The current is then only running in a circle and the solar panel can't output anything anymore.

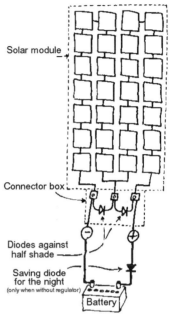

These diodes are there to prevent over-heating one half of the solar panel if it's half in the shade. The part which is still in the full sun would try to push current through the cells in the shade. When you set up your solar panels try to make sure this can't happen and they will run ok, also when you take these diodes out. But if you connect them the wrong way round again, you will blow up the inner connections of the solar panel, which usually will be irrepairable.

So it's better to replace these diodes (e.g. for a 50W panel you can use diodes like BY550-50 or P600A) and use a proper regulator, then this will never be a problem.

Broken Glass?

Solar panels with broken glass still give out up to 70% of their original power. If you recover it with another layer of crystal or plastic glass, you can save them from being killed by water and humidity. The glass will take another 10% away but if you don't recover it water would destroy the silicon cells and the contacts between them.

The best way is to glue a new layer of thick glass (5mm or 6mm) with polyurethane glue (like sikaflex never use silicon!) onto the aluminium frame. Make sure it's totally waterproof. You have to hold the glass with metal brackets on all sides, so that if the glue fails the glass plate can't fall from the roof!

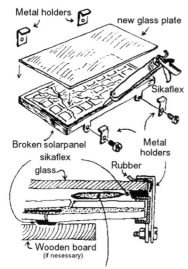

Tip: for condensing water between the old and the new glassplates put a small bag of drying stuff (out of the lid of a Vitamin C box) on a place where it doesn't cover the solar cells.

If the solar panel is bent and out of shape you can carefully fix a wooden plate onto the back to pull it back into shape.

Broken Connections?
The connector box with the cables and the solar panel can get ripped apart, usually when high winds have torn the panel from its mountings. The back of a solar panel is usually made of a flexible white PVC. This can be cut with a sharp knife, to get to the silver tracks inside. Solder new cables onto the silver tracks (you can do this with normal soldering stuff) and close the cut properly with some two component glue or a hot glue gun. Fix the box onto the frame again or glue it back on the back of the solar panel.

Old AEG Solar Panels
The old AEG solar panel series from the '80s was made with 40 cells (4 lines of 10), a thin stainless steel frame and crystal glass on both sides. These were connected with aluminium instead of silver tracks and many of them now have connection problems between the cells.

To repair them you have to grind away the glass from the back with a diamond disk, until you can solder some new cable connections onto the aluminium tracks. To find out where the bad connections are, press onto different parts of the solar panel when it's in full sun and measure the current coming out.

You need special aluminium soldering stuff and a powerful soldering iron (60W). Bridge the cells with bad connections using copper cables and solder into place.

This will result in less tension (-0.55V for each cell you bridge), but down to 16V it will still be able to charge a battery. Close the gaps with two component glue or hot glue. But also this will not be a final solution, because the other aluminium strips will corrode later.

Soldering On Solar Cells

The metal strips printed on solar cells are very thin and breakable. They melt easily and don't connect well with normal soldering stuff. SMD (Surface Mounted Device) soldering stuff which contains silver works for

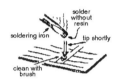 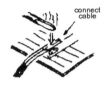

these very small electronic components. It's best to use very little of it and only make small connection points. Clean up afterwards using a brush and some water.

Soldering Irons

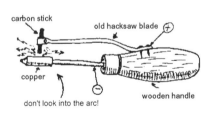

Professional soldering equipment with temperature controls usually work on 12V (though some are 24V). You can sometimes buy the soldering irons separately, they are usually not too expensive. If you get one, you can run it directly from your solar system.

A strong arc soldering iron can be made with a piece of carbon from an old 1.5V zinc carbon battery, fixed above a piece of copper on the tip of a home made iron. When the iron is not in use, the carbon has no contact with the copper. To heat up the iron, you press the carbon and copper together until they create an arc between them. This uses a lot of power but results in the copper tip becoming extremely hot.

12V heating coils with 20-80W of power can also be totally home-made. Use heating wire out of a 230V fan heater. Use a multimeter to measure the right length to reach the power you want (e.g. for

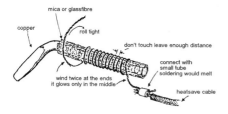

30W = 2.5A). If the heating cable is too thin it will glow bright red. In this case twist two of them together and measure the length again.

12V immersion heaters can be used to do something useful with the excess solar electricity, when the batteries are full. How about a solar tea or coffee? 50W to 100W boils a cup of water in just a few minutes! You can make immersion heaters by using a thin brass tube (from an old lamp), close the bottom by soldering a round piece of brass inside (use hard soldering stuff at 750°). Cover

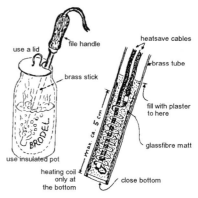

it inside with glass fibre mats, make a heating coil like the one for a soldering iron (see above) connect with heat-safe cables and stick it in. Fill the tube up with fire cement or plaster. Fitting a timer prevents you boiling away all the water if you forget to switch it off.

Food Mixers

Use an old whisk from a 230V food mixer, attached to a cordless drill to whip cream or mix a cake mixture. Squash the wire ends of the whisk towards the handle to make it shorter and wider, as this makes it more powerful.

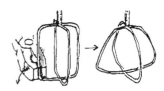

For hand blenders change the motor to a 12V windscreen wiper motor with a filed down axle (see the chapter, Solar Grinders, page 90). Use a piece of tube for the connection to the cutter from an old 230V blender.

Cooling Fans

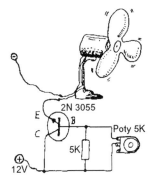

This also uses a windscreen wiper motor without its gear system; and an old fan, or a homemade fan with 3 or 4 rubber wings made from a truck tyre inner tube. This will make a strong fan which can run on the surplus electricity produced in the summer. You can even have 2 or 3 speeds from a windscreen wiper motor. The motors from car heating systems work well, too. With a poty and a big transistor (2N 3055) you can make a good speed control. Connect the poty with a fixed resistor (for minimal turning rate) onto the basis of the transistor. These fans will also work directly connected to a solar panel.

Slide Projectors

Projectors without cooling motors are the easiest to convert to 12V. Just put a halogen headlight bulb (55W or up to 100W if you use the not legal for road use ones) from a car into the projector.

Airbrush Systems

Using a 12V car air compressor and an old fire extinguisher, you can make a small pressurized air system. Empty the extinguisher fully, then unscrew the top. Pour out the remaining powder and remove the gas cylinder. To fill the extinguisher make a small hole for the input, and insert a tubeless tyre valve from a car. Connect the output of the compressor here and connect your airbrush to the output of the extinguisher. For a small spray system you could connect the airbrush directly to a truck horn compressor.

Timers

For electronics experts only! Digital timer switches designed for use with 230V often have 12V compatible electronics. You just have to exchange the 230V relay for a 12V relay

Sewing Machines

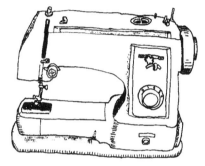

A cordless drill with an electronic speed governor can run a sewing machine. Take the trigger switch out of the drill and put it into the machine's foot pedal. Enlarge the cable, using appropriately thick wiring. You can either swap the 12V and the 230V motors directly, or connect the rotating chuck part of the drill to the wheel on the main axle of the sewing machine. A halogen lamp socket can be fitted into the E14 lamp socket on the machine. Use a washer and small screws to fix it together (see the chapter, Lamps page 86). A 5W halogen bulb gives enough light for sewing.

Relays

Economy Drive
Relays need a high current to switch them on initially, but once the contacts are together, it takes just a small percentage of the power to hold them there. Relays can be made more economic with a simple circuit; you just need a condenser and a resistor. When you switch the relay on the electricity flows over the condenser for a short time, giving a high current

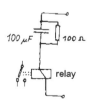

to switch the relay, but this current is cut off when the condenser is at full load. Then a smaller current flows over the resistor and holds the contacts in place. The size of the condenser and resistor differ for each relay, so you must experiment to see which is most suitable. You can easily cut your relay's power consumption by 50%.

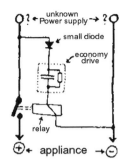

Polarity Safeguard

For a strong appliance you can make a polarity safeguard without the loss of tension associated with diodes. Only the coil of the relay gets power via the diode and so the relay will only switch the power on to the appliance when the polarity is correctly connected.

Start Current Safeguard

Big electric motors need a tremendous amount of current to start them. In the start phase the motor has hardly any resistance, so it is almost making a short circuit. A lot of the current is wasted in heat. Only when it runs faster does the resistance inside the motor increase to its normal rate. If you connect a power resistor, made of old heating cables or thin fence wire, in series to the motor it will start more softly. When the motor reaches a certain speed the tension at the motor rises and can now switch a relay on to shortcut the power resistor. The motor then recieves the full tension. To be able to adjust this point exactly, you can use a adaption resistor in front of the relay coil.

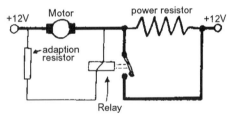

Rechargeable Torches

For this you can use any normal torch but with NiCd or NiMh batteries. With an input plug and a simple charging circuit placed somewhere in the torch you can charge them up on the 12V system. See the next section, Charging Batteries.

Charging Batteries

With a 12V system, you can charge up to six NiCd or NiMh rechargeable batteries (7.2V) at the same time. If you want to be able to charge constantly and not have to worry about exact charging times and the

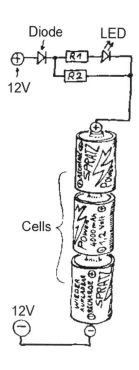

Diode LED

12V

Cells

12V

risk of overloading, you have to reduce the charging current to 1/20 or 1/30 of the capacity (a 600mAh rechargeable battery would then charge with 20mA). When you multiply the capacity (mAh) by 0.05 in a calculator, you will get out the amount in milliamps. See also page 30.

With a diode, two resistors and a light emitting diode (LED), the whole system becomes idiot proof. The LED shows if the batteries are well connected. The resistor varies according to the size and number of batteries you want to charge. See table below for some examples. But you also can use 2 potys (adjustable resistors) and adjust the currents to find out the correct resistors required. Pass 10mA through the LED with resistor 1 and the rest of the current through resistor 2.

No. of cells	Volts	LED 10mA R1	500 mAh R2 – 15mA	1300 mAh R2 – 50mA	4000 mAh R2 – 190mA
1	1.2V	860Ω	670Ω	190Ω	53Ω
2	2.4V	740Ω	600Ω	160Ω	47Ω
3	3.6V	620Ω	500Ω	140Ω	40Ω
4	4.8V	500Ω	430Ω	120Ω	34Ω
5	6.0V	380Ω	350Ω	100Ω	28Ω
6	7.2V	260Ω	270Ω	75Ω	21Ω

If you want to charge more than 6 cells at the same time, you need a higher tension than 12V. But you can do it with a bit of a trick. In the chapter, Bench Power Supplies, page 118, you will find an electronic circuit with a TDA 2003 to double the tension. The second part of the circuit with the LM 317 isn't needed. You can charge up packs of fifteen cells (18V) with this.

LED Lights

These new super bright LED lights that you get in expensive torches, are very economical and efficient. But it's cheaper to make them yourself. Connect three NiCd or NiMh batteries in series and put a resistor of 10Ω to 20Ω inbetween to cut down the current to about 20mA.

Music Systems

The Mini-Marshall Amp

The Mini-Marshall is a cheap battery powered guitar amp which also runs with 12V. You can connect the amp to a much more powerful speaker. Make a new jack socket in the amp, to connect to the speaker. This should include a switch to change between the amp's built-in speaker and the external one. Don't connect a guitar speaker with a resistance of less than 8Ω but the size of the speakers doesn't matter at all. You will be amazed what a noise this small amp of 0.8W makes when it is connected to a 4x 12in Marshall speaker cabinet.

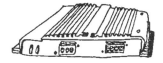

Amplifiers

Big car amplifiers run directly with 12V. You can find them in all classes of power 50W to 3,000W. The sound quality is mostly comparable to professional 230V power amplifiers. With 12V pre-amps you can make guitar, bass and PA amplifiers. You can buy pre-amps kits from electronic shops but if you want to make your own this is a circuit for an easily made pre-amp. You can amplify both condenser and dynamic microphones with it. Plug the pre-amp directly into the line in of the power amplifier.

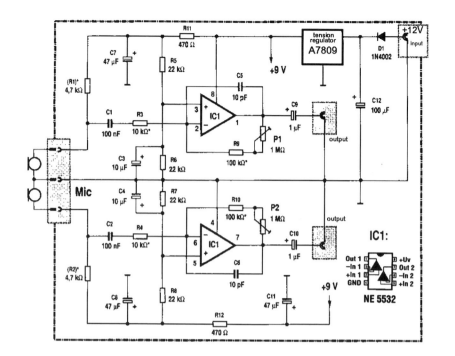

Microphone pre-amplifier circuit.

4-Track Recorders

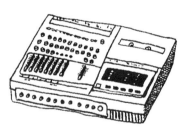

Many middle of the range 4-track recorders have an external 12V transformer and can be used with a 12V system. But be careful, sometimes you cannot connect the minus (or earth) of the in and outputs with the minus of the battery, so it's safer to have an extra battery, exclusively for the 4-track, or at least for the external effects and pre-amps.

Washing Laundry

It is possible to run a washing machine on 12V. But to keep it fully automatic you have to change all the 230V motors for new 12V motors. Only the spinning and the electrical heating can't be done easily like this. But you can use warm water from a solar collector, and an old spin dryer also adapted to 12V(see below). You can adapt a windscreen wiper motor to the back of the cylinder of an old washing machine by using a bicycle chain and cycle cogs. With some luck the whole pedal will fit directly onto the axle of the washing machine, but if not you can screw the big bicycle cog (well centred) onto the aluminium wheel.

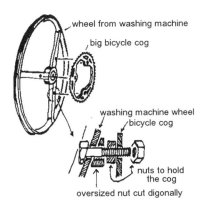

wheel from washing machine

big bicycle cog

washing machine wheel
bicycle cog

nuts to hold the cog

oversized nut cut digonally

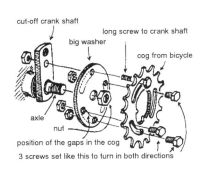

cut-off crank shaft

long screw to crank shaft

big washer

cog from bicycle

axle

nut

position of the gaps in the cog
3 screws set like this to turn in both directions

Fix the small bicycle cog onto the windscreen wiper motor use a big washer and drill a new hole into the crank shaft.

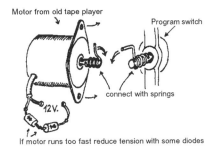

Motor from old tape player

Program switch

connect with springs

12 V.

If motor runs too fast reduce tension with some diodes

To change the program switch you need to fix a small motor from an old tape player instead of the 230V motor. You also need to convert the water pump by connecting a car heater fan motor in a similar way to the dryer (see spin dryer instructions).

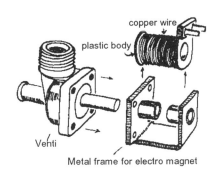

copper wire

plastic body

Venti

Metal frame for electro magnet

You have to rewind the valves which let the water in with thicker copper cable (0.15mmØ). If this is too complicated for you, you can also do the water in and out by hand.

Also you can't connect the windscreen wiper motor and the new pump motor directly to the program switch, because these 12V motors use much more current and the contacts inside the program switcher would burn out. Use old car relays to switch them instead.

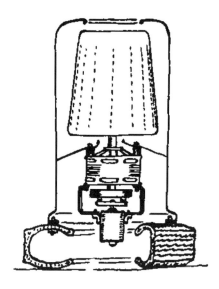

For spinning use an old spin dryer. You can't take out the old motor because it includes all the bearings. However you can adapt a strong car radiator fan motor so that it turns the old motor. It is important that you use a start current safeguard relay circuit (see page 111) otherwise the motor will get very hot. The new motor sticks out at the bottom, but this doesn't matter if you fix the whole thing onto an old car wheel. The spin dryer now uses 20 per cent the electricity of the old 230V one.

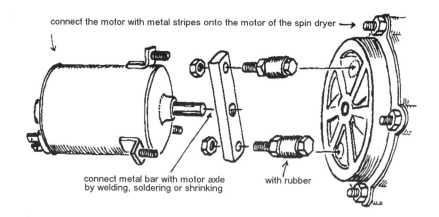

connect the motor with metal stripes onto the motor of the spin dryer →

connect metal bar with motor axle by welding, soldering or shrinking

with rubber

Connection detail to fit a car cooling fan motor to existing spin drier motor.

Bench Power Supply

With a simple circuit made with a TDA 2003 you can double the tension of a 12V system. After the tension has been doubled you can use a voltage regulator to constantly regulate the voltage between 1.25V and 24V. With this set up you can test solar regulators, deep discharging regulators and the economical relay circuits at different tensions. Don't forget to connect the TDA2003 and the LM317 to a heat sink. They need separate heat sinks and they have to be well insulated from each other.

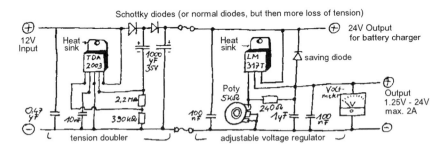

Battery Charger

If you want to charge NiCd or NiMh batteries, you can connect a charger to the 24V output of the bench power supply and charge power packs of 1 to 15 cells (1.2V to 18V).

With a 10 position switch, you switch on the right charging current. The transistor (BC557) is only for the LED and is not strictly necessary. The 7805 can be connected to the same heat sink as the TDA2003. The 10Ω resistor must have a minimum of 5W.

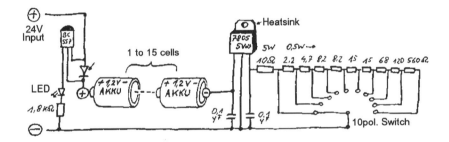

118

Home-made Regulators

You can make charging regulators and deep discharging regulators very simply using common components, possibly scounged from old electronic appliances. They need very little electricity and make minimal electrosmog. Using old car relays, you can control currents of up to 30A without problems. These circuits are very sensitive to dirt and damp, and should be coated with paint to protect them.

Solar Regulator

The poty (potentiometer) of 10kΩ is to switch on the relay and the poty of 250kΩ is to switch it off.

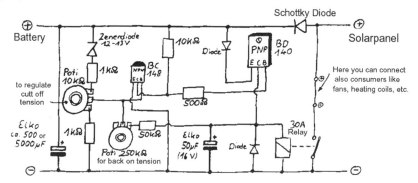

Deep Discharging Regulator

When the battery is empty the relay switches off the appliance and switches on an alarm system, e.g. from an old digital clock.

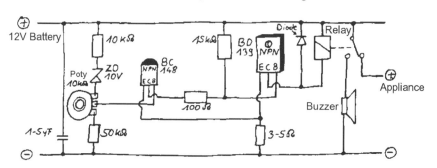

Here are two rather more modern circuits made with Mosfets.

Solar Regulator

The solar regulator switches off the solar panel by creating a short circuit (shunt) to the solar panel for about 30 seconds. So this regulator is nearly electrosmog free.

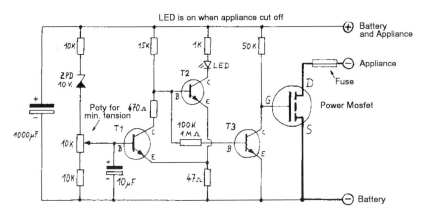

Deep Discharging Regulator

With it's LED and 47Ω resistor it switches back on again when the tension is 0.5V higher than the settled cut off point. By using more Mosfets in parallel you can switch any amount of current you like.

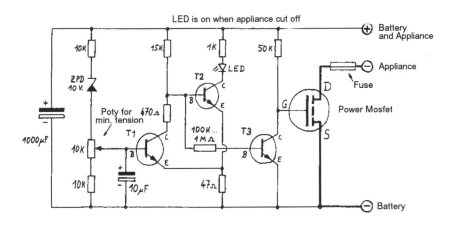

Tables

CONDENSER CODES

pF	nF	μF	Code	e.g.
10	0,01	0,00001	xx0	100
100	0,1	0,0001	xx1	101
1000	1	0,001	xx2	102
10000	10	0,01	xx3	103
100000	100	0,1	xx4	104
1000000	1000	1	—	—
10000000	10000	10	—	—

Example Condensers

104 = 100nF

222 = 2.2nF

473 = 47nF

n10 = 100pF

2200 = 0.0022μF

10p = 10pF

153 = 15μF

Example Resistors

red red yellow 220kΩ

yellow violet red 47kΩ

yellow violet brown 470kΩ

brown black green 1MΩ

blue grey black 68Ω

red red gold 22Ω

brown black black orange   100kΩ

orange orange black orange 330kΩ

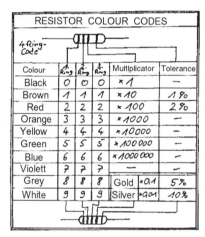

RESISTOR COLOUR CODES

4-Ring-Code

Colour	1.Ring	2.Ring	3.Ring	Multiplicator	Tolerance
Black	0	0	0	x 1	—
Brown	1	1	1	x 10	1 %
Red	2	2	2	x 100	2 %
Orange	3	3	3	x 1000	—
Yellow	4	4	4	x 10000	—
Green	5	5	5	x 100 000	—
Blue	6	6	6	x 1000 000	—
Violett	7	7	7	—	—
Grey	8	8	8	Gold x0,1	5%
White	9	9	9	Silver x0,01	10%

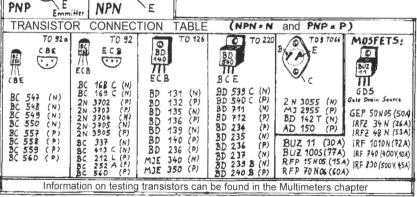

Basis | Collector C
B — PNP — E Emmitter
B — NPN — C / E

TRANSISTOR CONNECTION TABLE (NPN=N and PNP=P)

TO 92a   CBE

TO 92   ECB   CBE   ECB

BC 168 C (N)
BC 169 C (N)
2N 3702 (P)
2N 3703 (P)
2N 3704 (N)
2N 3705 (N)
2N 3905 (P)

BC 547 (N)
BC 548 (N)
BC 549 (N)
BC 550 (N)
BC 557 (P)
BC 558 (P)
BC 559 (P)
BC 560 (P)

BC 337 (N)
BC 413 C (N)
BC 212 L (P)
BC 252 A (P)
BC 560 (P)

TO 126   ECB

BD 131 (N)
BD 132 (P)
BD 135 (N)
BD 136 (P)
BD 139 (N)
BD 140 (P)
BD 236 (P)
MJE 340 (N)
MJE 350 (P)

TO 220   BCE

BD 539 C (N)
BD 540 C (P)
BD 711 (N)
BD 712 (P)
BD 234 (P)
BD 235 (P)
BD 236 (P)
BD 237 (N)
BD 233 B (N)
BD 240 B (P)

TO3 TO66   C

2 N 3055 (N)
MJ 2955 (P)
BD 142 T (N)
AD 150 (P)

BUZ 11 (30A)
BUZ 100S(77A)
RFP 15 N05 (15A)
RFP 70 N0% (60A)

MOSFETS:

GDS
Gate Drain Source

GEP 50N05 (50A)
IRFZ 34 N (26A)
IRFZ 48 N (53A)
IRF 1010N(72A)
IRF 740 (400V,10A)
IRF 830 (500V, 4.5A)

Information on testing transistors can be found in the Multimeters chapter

Further Reading

Permaculture Magazine – Solutions For Sustainable Living
Quarterly colour magazine, packed with inspirational ideas and informative articles, designs, news, book, product and tool reviews, letters, classified ads and details of permaculture and special courses. Incorporates Global Ecovillage Network News. 68pp.

Practical Photovltaics, Richard J Kemp, Aatec Publications, 1995.

Self Reliance, John Yeoman, Permanent Publications, 2000.

Other Titles in the Simple Living Series

Building A Low Impact Roundhouse, Tony Wrench, Permanent Publications, 2001.

Eat More Raw – A Guide To Health And Sustainability, Steve Charter, Permanent Publications, 2004.

Getting Started In Permaculture – Over 50 Projects For House And Garden Using Recycled Materials, Ross and Jenny Mars, Permanent Publications, 2007.

Tipi Living, Patrick Whitefield, Permanent Publications, 2000.

These titles, and over 500 others relating to sustainable living, plus useful tools and products are available from:

THE GREEN SHOPPING CATALOGUE
Permanent Publications, The Sustainability Centre, East Meon, Hampshire GU32 1HR, U.K. Tel: 0845 458 4150 (local rate UK only) or 01730 823 311
Overseas +44 1730 823 311. Email: info@green-shopping.co.uk
Order online at: www.green-shopping.co.uk

Notes

Notes